Total Productive Maintenance
in America

Total Productive Maintenance
in America

Published by the
Society of Manufacturing Engineers
Dearborn, Michigan

Copyright © 1995 by Society of Manufacturing Engineers

9876543

All rights reserved, including those of translation. This book, or parts thereof, may not be reproduced in any form or by any means, including photocopying, recording, or microfilming, or by any information storage and retrieval system, without permission in writing of the copyright owners.

No liability is assumed by the publisher with respect to the use of information contained herein. While every precaution has been taken in the preparation of this book, the publisher assumes no responsibility for errors or omissions. Publication of any data in this book does not constitute a recommendation or endorsement of any patent, proprietary right, or product that may be involved.

Library of Congress Catalog Card Number: 95-067392
International Standard Book Number: 0-87263-461-2

Additional copies may be obtained by contacting:

Society of Manufacturing Engineers
Customer Service
One SME Drive
Dearborn, Michigan 48121
1-800-733-4763

SME staff who participated in producing this book:

Donald A. Peterson, Senior Editor
Rosemary K. Csizmadia, Operations Administrator
Sandra J. Suggs, Editorial Assistant
Dorothy M. Wylo, Production Assistant
Cover design by Judy D. Munro, Manager, Graphic Services

In charge of video production:

William F. Masterson, Producer

Printed in the United States of America  Printed on recycled paper

Table of Contents

Foreword vii

Preface ix

Chapter 1 — The Foundation for World-class Manufacturing 1

Chapter 2 — Elements of Successful TPM 17

Chapter 3 — Getting TPM Going 23

Chapter 4 — Building TPM Teams 35

Chapter 5 — Learning the Ropes and Staying Abreast 41

Chapter 6 — TPM in a Union Environment 47

Chapter 7 — Predicting the Inevitable and
Preventing the Exceptional 61

Chapter 8 — CMMS—The Common Denominator 79

Chapter 9 — The Equipment Operator as Equipment Owner 93

Chapter 10 — Putting it all Together 101

Bibliography 111

Index 113

Foreword

WHAT IS TPM ANYWAY?

Travelling to different companies and meeting with people from many different types of plants, I continue to hear the question: "Why TPM? What is it about TPM that's so much better than all of the other stuff we've got going on anyway?" Mostly, those who ask this question are not that familiar with what "TPM" really is and what it can accomplish. What *is* TPM anyway?

First, TPM—total productive maintenance or total productive manufacturing, as more and more American companies prefer to call it—is a way of working together to improve equipment effectiveness in modern manufacturing plants and commercial facilities. TPM means that *everybody* who works on or around the equipment, not just the maintenance department, is looking for ways to keep the equipment running when it should, as fast as it should, with the highest levels of quality and yield possible. TPM is about people working together to make sure the equipment is operated and maintained the way it should be and to make sure that newly designed equipment is easier to install, start up, operate, and maintain than that which it replaces.

Second, TPM requires a certain discipline about the fundamental ways people care for their equipment. Yes, *their* equipment. In a TPM work culture there is a sense of ownership that I typically do not see in most plants in the United States: people treating their equipment as if it were their own car or truck. This means paying attention to the funny noises it makes, or the vibrations, or the leaks, or the smoke coming from the motors. It means keeping it clean so they can see problems before they become failures. The required discipline of TPM refers to the way *everybody* who operates, maintains, monitors, and supports the equipment cares for it. This includes performing appropriate maintenance inspections, adjustments, and lubrication, and keeping detailed records about the equipment performance and care. Discipline also refers to the training programs, materials, and methods used to ensure that everyone is qualified to perform every aspect of their job roles safely and effectively.

Third, TPM requires that operators become trained to perform some tasks that are traditionally thought of as *maintenance* work. But then, is it really maintenance work? Being trained by skilled maintenance people in your plant to do routine preventive maintenance (PM) inspections, adjustments, and minor lubrication is not really that difficult. In fact, most operators I know perform similar tasks on their own cars and trucks and around the home or farm. It saves them money and keeps their equipment running. Like most of us, they also know when the task exceeds their skill level and knowledge and that it is then time to call in a professional maintenance technician.

And, finally, TPM is not a single thing, program, or tool. TPM is all of what I have just discussed: *working together to improve equipment effectiveness more than each individual activity or program could possibly achieve by itself.* It's called *synergy* and *interdependence*. That's TPM. Of course, working together in this way may challenge the way people have historically thought about work and their jobs. I would estimate that clearly 80 percent of the success of TPM will be based on how the "people" issues are addressed in your company. TPM is not as much technology and tools as it is mindsets, paradigms, habits, knowledge, skills, and beliefs. The "soft stuff."

This book and the accompanying video series contain many insights into how companies, individuals, and work groups are making TPM work for them. In our western culture, there are no real "TPM cookbooks" that outline each and every step to transform your company and the habits of those who operate, maintain, and design equipment. What works for one company will not necessarily succeed in another. That is the nature of our diverse, and often unique, work cultures and work systems. But, we *can* learn from each other. We can learn how others think about and establish their TPM work cultures in ways that give our companies new foundations from which to build more effective manufacturing and support equipment.

The starting points for TPM *will* vary from company to company, plant to plant, and quite often from department to department depending on the needs of the equipment and of the people. Each of the companies featured in this book and the video series selected their starting point for TPM based on the unique aspects of their work cultures and business needs. Their stories give us a rich perspective of many of the elements and concepts of TPM. The more all of the elements and concepts of TPM are put to work, the more effective your equipment will be, both short-term and over the long haul. No doubt about it! So, start now—put the resources behind your TPM initiatives and measure the results.

Robert M. Williamson, 1995
Strategic Work Systems[SM]

Preface

The importance of low inventory operating policies, such as JIT, and the emphasis on quality assurance programs, such as TQM, are well known and have been widely adopted by manufacturers competing in world markets. The realization that productivity, inventory, safety and health, and quality all depend on *equipment performance* is a more recent phenomenon which has piqued interest in companies worldwide.

Performance of facilities and equipment is critical to a manufacturer's ability to produce low-cost, high-quality products; thus, effective equipment maintenance has become increasingly important in the manufacture of quality products. This recognition of equipment's role in manufacturing has led to the development and growing implementation of a comprehensive concept of equipment repair, service, maintenance, and management called total productive maintenance (TPM).

TPM is a methodology and philosophy of strategic equipment management focused on the goal of building product quality by maximizing equipment effectiveness. It embraces the concepts of continuous improvement and total participation by all employees and by all departments. It targets elimination of waste caused by equipment downtime through autonomous group activities and individual operator involvement in tasks found conventionally in the maintenance department.

Total Productive Maintenance in America is an instructional package of five videocassettes and this reference book produced by the Society of Manufacturing Engineers. It tells the TPM story through the experiences of five major U.S. companies that have pioneered TPM in the U.S.: Harley-Davidson, Magnavox, Briggs & Stratton, AT&T, and The Timken Company.

Content of the book parallels that of the video series and was written to serve as a guide for manufacturers as they prepare their competitive strategy for the twenty-first century. The first three chapters describe the hurdles to overcome in getting buy-in from management and shop-floor workers and the types of changes needed to successfully implement TPM.

Chapters 4 and 5 focus on the teams needed to make TPM work and the training of team members for autonomous maintenance. Central to these are the concepts of empowerment and operator ownership, especially as they apply to keeping *overall equipment effectiveness* high.

In chapter 6, Kevin McGlynn of Briggs & Stratton recounts some of the unique considerations attending the introduction of TPM in a union shop and how the collaboration of his company and its organized labor force led to the successful implementation of TPM.

The importance to TPM of preventive and predictive maintenance is brought to the grass-roots level in chapter 7 as recommendations for developing objectives and a maintenance plan within the framework of a company's particular personality.

Noted TPM expert Terry Wireman describes CMMS, the automated dimension of TPM, in chapter 8. CMMS — the computerized maintenance management system — organizes all equipment information into a workable database that coordinates maintenance activities throughout an entire company and helps identify areas for improvement.

Chapter 9 describes the concept of autonomous maintenance, in which equipment maintenance tasks are done by a team comprised of those workers involved with the equipment on a daily basis—operators and maintenance tradespeople.

The last chapter of the book brings all the material together in a succinct and practical snapshot of what TPM is, what it does, and what it entails, written by TPM expert Robert Williamson.

Acknowledgments

The *Total Productive Maintenance in America* book and video series was, by design, an effort to gather, compile, and disseminate in the most comprehensible form current, accurate, and real-world information on TPM and how it is practiced by leading U.S. manufacturers. This endeavor would not have been possible without the extraordinary contributions of several people closely associated with TPM in America.

SME is indebted to the following people for their selfless participation in the preparation of material for the videos and this reference guide.

Our thanks to Terry Wireman, educator, adviser, and widely read author on TPM, for his intellectual energy in writing major portions of the book and reviewing the entire manuscript.

To Robert Williamson, internationally recognized consultant, writer, and educator for authoring the Foreword and closing chapter of the book. His editorial contributions and critical review of the manuscript and resulting recommendations were invaluable in helping to make the work an authoritative text.

To Edward Stanek, equipment maintenance expert and consultant to industry, for applying his many years of experience in maintenance systems design and technology to ensuring the accuracy and integrity of the content of both the video and editorial portions of the series.

To Kevin McGlynn of Briggs & Stratton for his insights into implementing TPM in a union environment, as described in chapter 6.

And to those in our featured companies who gave of their time in interviews, discussions, and on-camera appearances:

AT&T
Robert Archer
Anthony Cappabianca
Horace DesRochers
Raymond Donahue
Don Dooley
Richard Esposito
David Ierardi
Joe Kanan
Brian Murphy
Theresa Yau

Briggs & Stratton
James Benoit
Bob Blau
Dennis Erickson
Richard Hanson
John Heim, Sr.
Warren Kloth
Kevin McGlynn
Bob Phelps
Larry Schweitzer

Harley-Davidson
Patrick Bonnier
Arthur Koltermann
Dan Scherbert
Fredric Strege
Wayne Vaughn

Magnavox Electronic Systems
Sherry Baldridge
Richard Burdek
Paula Fritz
Curtis Hapner
Denise Kendrick
James Rice, Jr.
Thomas Rife
James Thompson
Phillip Wolfe
Audrey Woollweeven

The Timken Company
Mike Blackwell
Kenneth Fowler
Michael Segina
William Spake II
Perry Stephens
Dennis Wilkins

CHAPTER 1

The Foundation for World-class Manufacturing

TPM is important and necessary in manufacturing operations because it provides the capability to take the talents of people and educate them in ways so that benefits can be derived from their knowledge of equipment and machinery. We make process improvements, we make equipment improvements, we make quality improvements because of the ideas of our people. TPM is one way that we've used to educate our people in a broad spectrum of issues related to running an operation.

—Mike Blackwell, Manager, Green Machining and Heat Treat Operations, The Timken Company, Gaffney Bearing Plant

Good maintenance is good business. The prime motivator in manufacturing, especially as it pertains to equipment maintenance, is to keep production running in high gear. Competition mandates it. Maintenance directly affects the productivity, quality, and direct costs of production. Yet today, the most-practiced approach to maintenance—that of focusing on equipment only when it breaks down—stands in direct opposition to the target of high productivity, with the postmortem being that production stops and the maintenance department draws exceptional (and unwanted) visibility owing to the extraordinary costs such practices incur in terms of competitiveness and real dollars.

WHAT IT IS

To keep production in high gear—in fact, to survive—manufacturers are increasingly obliged to move from a "panic" maintenance mindset toward a concept of "productive" maintenance, one organized around a well trained staff, within a carefully defined plan, and with meaningful participation of employees outside of what is normally thought of as maintenance. It's a move toward a total team approach of preventive maintenance (PM) and total quality management (TQM).

Total productive maintenance (TPM) is a concept evolved by the Japanese in the 1960s from maintenance practices they learned from the United States during the 1950s. Total productive maintenance is perhaps a misnomer for the concept, since it goes beyond just maintenance. TPM is an *equipment management strategy* that involves all hands in a plant or facility in equipment or asset utilization. TPM is to productivity as flexibility is to competitiveness. Without it, corporate survival is in jeopardy.

At the core of TPM is a new partnership among the manufacturing or production people, maintenance, engineering, and technical services to improve what is called overall equipment effectiveness (OEE). It is a program of zero breakdowns and zero defects aimed at improving or eliminating the six crippling shop-floor losses:

- Equipment breakdowns.
- Setup and adjustment slowdowns.
- Idling and short-term stoppages.
- Reduced capacity.
- Quality-related losses.
- Startup/restart losses.

A concise definition of TPM is elusive, but *improving equipment effectiveness* comes close. The partnership idea is what makes it work. In the Japanese model for TPM are five pillars that help define how people work together in this partnership.

Five Pillars of TPM

1. *Improving Equipment Effectiveness.* In other words, looking for the six big losses, finding out what causes your equipment to be ineffective, and making improvements.
2. *Involving Operators in Daily Maintenance.* This does not necessarily mean actually performing maintenance. In many successful TPM programs, operators do not have to actively perform maintenance. They are involved in the maintenance *activity*—in the plan, in the program, in the partnership, but not necessarily in the physical act of maintaining equipment.
3. *Improving Maintenance Efficiency and Effectiveness.* In most TPM plans, though, the operator is directly involved in some level of maintenance. This effort involves better planning and scheduling, better preventive maintenance, predictive maintenance, reliability centered maintenance, spare parts equipment stores, tool locations—the collective domain of the maintenance department and the maintenance technologies.

4. *Educating and Training.* This is perhaps the most important task in the TPM universe. It involves everyone in the company to varying degrees: training operators to operate machines properly and maintenance people to maintain them properly. If operators will be performing some of the preventive maintenance inspections, training involves teaching operators how to do those inspections and how to work with maintenance in a partnership. Also involved is training supervisors on how to supervise in a TPM-type team environment.
5. *Designing and Managing Equipment for Maintenance Prevention.* Equipment is costly and should be viewed as a productive asset for its entire life. Designing equipment that is easier to operate and maintain than previous designs is a fundamental part of TPM. Suggestions from operators and maintenance technicians help engineers design, specify, and procure more effective equipment. And, by evaluating the costs of operating and maintaining the new equipment throughout its life cycle, long-term costs will be minimized. Low purchase prices do not necessarily mean low life-cycle costs.

COMPETITIVE ADVANTAGE

In most companies today, management is looking for every possible competitive advantage. Companies focus on total quality programs, just-in-time (JIT) programs, and total employee involvement (TEI) programs. All require complete management commitment and support to be successful. Figure 1-1 depicts TPM's fit in the manufacturing mix. However, no matter how management works to make these programs produce results, it is impossible to be totally successful without integrating these with a TPM program.

This is a bold statement; however, it is impossible to make the programs mentioned function effectively without equipment or assets that are currently maintained in world-class condition. Consider:

- Is it possible to produce quality products on poorly maintained equipment?
- Can quality products come from equipment that is consistently out of specification or worn to the point that it cannot consistently hold tolerance?
- Can a JIT program work with equipment that is unreliable or has low availability?
- Can employee involvement programs work for long if management ignores the pleas to fix the equipment or get better equipment so a *world-class* product can be delivered to the customer on a timely basis, thus satisfying the employee concerns and suggestions?

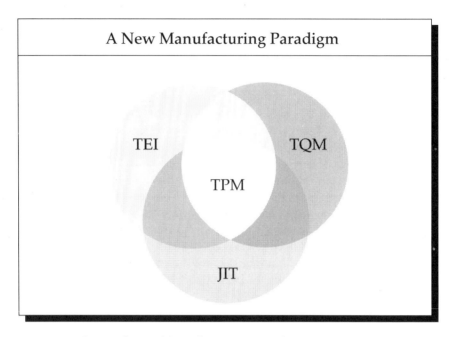

Figure 1-1. At the core of successful manufacturing, TPM catalyzes proven productivity programs.

TPM can improve reliability, maintainability, operability, and profitability—but achieving these goals requires the talents and involvement of every employee. Through autonomous activities—in which the operator is involved in the daily inspection and cleaning of his or her equipment—companies will discover the most important asset in achieving continuous improvement—its people.

Companies are beginning to realize that the management techniques and methods previously used to maintain equipment are no longer sufficient to compete in world markets. Attention is beginning to focus on the benefits of TPM, yet the number of companies successfully implementing TPM is relatively small. The reason is that many companies try to use TPM to compensate for an immature or dysfunctional maintenance operation. They fail to realize that TPM is an evolutionary step, not a revolutionary one. To fully understand the character of TPM, it is necessary to consider the evolution of a typical quality program.

The Quality Continuum

In Figure 1-2, the various stages of the continuum of a quality program are highlighted along the bottom of the arrow. In the early days, a company would ship almost anything to the customer. If the product did not meet

Figure 1-2. The various stages of the quality maturity continuum target a zero-defects quality goal.

customer standards, nothing was done about it until the customer complained and shipped it back. However, this eventually became costly when competitors would ship products that the customer would accept, since there was no quality problem. Complacency sabotaged competitiveness. To stay in the game, the company was forced to make changes in the way they did business.

The second step was to begin inspecting the product in the final production stage, or in shipping just before it was loaded for delivery to the customer. Though this was better than before (since it reduced the number of customer complaints) the company realized that it was expensive to produce a product only to reject it just before it was shipped. In effect, they were shooting themselves in the foot. It was far more economical to find the defect earlier in the process and eliminate running defective material through the rest of the production process. Common sense.

This led to the third step to quality system maturity, the development of the quality department. This department's responsibility was to monitor, test, and report on the quality of the product as it passed through the plant. At first blush, this seemed to be much more effective than before, with the defects being found earlier, even to the point of statistical techniques being used to anticipate or predict when quality would be out of limits. However, there were still problems. The more samples the quality department was required to test, the longer it would take to get the results back to the operations department. It was still possible to produce minutes', hours', or even shifts' worth of product that was defective or out of tolerance before anyone reined in the errant piece of equipment.

Solving this problem led to the fourth step, training the operators in the statistical techniques necessary to *monitor* and *trend* their own quality. In this way, the term "quality at the source" was coined. This step enabled the operator to know down to the individual part when it was out of tolerance—and no further defective components were produced. This eliminated the production of any more defects and prevented rework and expensive downstream scrap. However, there were still circumstances beyond the control of the operator that contributed to quality problems, which led to the next step, the involvement of all departments of the company in the quality program. From the product design phase, through the purchasing of raw materials, to final production and shipping of the product, all involved recognized that producing a quality product for the lowest price, the highest quality, and the quickest delivery was the goal of the company. This meant that products were designed for producibility, the materials used to make the product had to be of the highest quality, and the production process had to be closely monitored to ensure that the final product was of perfect quality. The company had evolved to the stage of maturity necessary to be world class.

The Maintenance Continuum

How does this path to maturity relate to the path to maturity for asset or equipment maintenance? Figure 1-3 compares the two. In stage 1 of the path to TPM, the equipment is not maintained or repaired unless the customer (operations, production, or facilities) complains that it is broken. Only then will the maintenance organization work (or in some cases be allowed to work) on the equipment. In other words, "if it ain't broke, don't fix it." However, over time companies began to realize that when equipment breaks down it always costs more and takes longer to fix than if it was maintained on a regularly scheduled basis. This cost is compounded when the actual cost of the downtime is calculated. Companies began to question the policy, coming to understand that it is cost effective to allow the equipment to be shut down for shorter periods of time for minor service to reduce the frequency and duration of the breakdowns.

This leads to the second step on the road to TPM, the establishment of a good preventive maintenance program or building on one already in

The Foundation for World-class Manufacturing

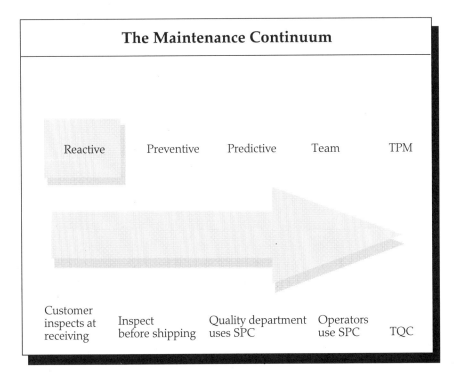

Figure 1-3. Total quality management coupled with total productive maintenance produces a zero-defect, zero-breakdown production environment.

place. This step allows for the inspection and routine servicing of the equipment before it fails and results in fewer breakdowns and equipment failures. In effect, the product is inspected before the "customer" gets it. Some of the techniques of PM include routine lubrication and inspections for major defects.

This second step, while producing some results, is not sufficient to prevent certain types of failures. The third step, then, is to implement predictive and statistical techniques for monitoring the equipment. The most common of these techniques are:
- Vibration analysis.
- Spectrographic oil analysis.
- Thermographic or infrared temperature monitoring.
- Nondestructive testing.
- Sonics.

The information produced from proper utilization of these techniques reduces the number of breakdowns to a low level, with overall availability being more than 90 percent. At this point, the "hidden" problems are discovered before they develop into major problems. However, the quest for continuous improvement emphasizes the need to do better. This leads to the fourth step: involvement of the operators in maintenance activities.

This step does not mean that all maintenance activities are turned over to the operators. Only the basic tasks are included, such as some inspection, the basic lubrication, adjusting, and routine cleaning of the equipment. The rationale for having operators involved in these activities is that they best know when something is not right with the equipment. In actual practice the tasks they take over are the ones that the maintenance technicians have trouble finding the time to do. Freed of the burden of doing some of the more routine tasks, the maintenance technicians can concentrate on refining the predictive monitoring and trending of the equipment. They also will have more time to concentrate on equipment failure analysis, which will prevent future or repetitive problems on the equipment. This step increases not only the availability of the equipment, but also its reliability over its useful life.

The last step of the evolution process is involving all employees in solving equipment problems, thereby increasing equipment effectiveness. The most common method is the use of cross-functional teams formed of members from various organizational disciplines to produce total solutions for these problems. Through team-building training, the team members learn the function, need, and importance of each team member, and in a spirit of understanding and cooperation, allow for production and service to reach world-class standards.

To reach these goals, certain resources must be in place or accounted for. They can be divided into three main categories. If not in place, they become obstacles to achieving the goals of TPM.

- Management support and understanding.
- Sufficient training.
- Allowance for sufficient time for the evolution.

Management support is essential. Management must completely understand the true goal of the program and back it. If management begins the program by emphasizing its desire to eliminate maintenance technicians, they have failed to understand the program's true purpose. The real goal is to increase overall equipment effectiveness, not reduce the labor head-count. Without management understanding of the true goal of asset utilization, the program is doomed to failure.

Sufficient skills training is another of the major activities of TPM. It must be given to at least two different levels. The first addresses the increased skills required for the maintenance technicians. The technicians will be trained in advanced maintenance techniques, such as predictive maintenance and equipment improvement. They also must have extensive training and guidance in data analysis to prepare them to find and solve equipment failure and effectiveness problems. Refresher training in the fundamentals of sound equipment maintenance methods is also considered a vital part of TPM.

Second is operator training. The operators must be trained to do basic maintenance on their equipment. In areas such as inspections, adjustments, bolt tightening, lubrication, and proper cleaning techniques, operators must be taught in detail. Also, before doing any repairs, operators must receive training to be certified to do the assigned tasks. Without proper training in selected skills, the equipment's effectiveness will decrease. The degree of operator involvement must also fit with the company culture.

Additional training for work groups, leadership, engineers, planners, and others is a vital part of the TPM work culture.

The last element is allowing enough time for the TPM evolution to occur. The change from a reactive program to a proactive program will take time. By some estimates it may be a 3- to 5-year program to achieve a competitive position. By failing to understand this point, many managers condemn their programs to failure before they ever get started.

Successful TPM programs focus on specific goals and objectives. When the entire organization understands the goals and how they affect the competitiveness of the company, the company will be successful.

The five central objectives of TPM are to:
1. Ensure equipment capacities.
2. Develop a program of maintenance for the entire life of the equipment.
3. Require support from all departments involved in the use of the equipment or facility.
4. Solicit input from all employees at all levels of the company.
5. Use consolidated teams for continuous improvement.

Ensuring equipment capacity emphasizes that the equipment performs to specifications. It operates at its design speed, produces at the design rate, and yields a quality product at these speeds and rates. The problem is that many companies do not even know the design speed or rate of production of their equipment. This allows management to set arbitrary production quotas. A second problem is that over time, small problems cause operators to change the rate at which they run equipment. As these problems continue to build, the equipment output may be only 50 percent of what it was designed to be. This will lead to the investment of additional capital in equipment, trying to meet the required production output.

Implementing a program of maintenance for the life of the equipment is analogous to the popular preventive and predictive maintenance programs companies presently use to maintain their equipment, but with a significant difference; it changes just as the equipment changes. All equipment requires different amounts of maintenance as it ages. A good preventive/predictive maintenance program takes these changing requirements into consideration. By monitoring failure records, trouble calls, and basic equipment conditions, the program is modified to meet the changing needs of the equipment.

A second difference is that TPM involves all employees, from shop floor to top floor. The operator may be required to perform basic inspecting, cleaning, and lubricating of the equipment, which is really the front-line defense against problems. Upper managers may be required to assure that maintenance gets enough time and budget to properly finish any service or repairs required to keep the equipment in condition so that it can run at design ratings.

Requiring the support of all departments involved in the use of the equipment or facility will ensure full cooperation and understanding of affected departments. For example, including maintenance in equipment design/purchase decisions ensures that equipment standardization will be considered. The issues surrounding this topic alone can contribute significant cost savings to the company. Standardization reduces inventory levels, training requirements, and startup times. Proper support from stores and purchasing can help reduce downtime, but more importantly it will aid in the optimization of spare parts inventory levels, thus reducing on-hand inventory.

Soliciting input from employees at all levels of the company provides employees with the ability to contribute to the process. In most companies this step takes the form of a suggestion program. However, it needs to go beyond that; it should include a management with no doors. This indicates that managers, from the front line to the top, must be open and available to listen to and consider employee suggestions. A step further is the response that should be given to each discussion. It is no longer sufficient to say "That won't work" or "We are not considering that now." To keep communication flowing freely, reasons must be given. It is just a matter of developing and using good communication and management skills. Without these skills, employee input will be destroyed at the outset and the ability to capitalize on the greatest savings generator in the company will be lost.

The creation of consolidated teams for continuous improvement begins with objective 4. The more open management is to the ideas of the work force, the easier it is for the teams to function. These teams can be formed by areas, departments, lines, process, or equipment. They will involve the operators, maintenance, and management personnel. They will, depending on the needs, involve other personnel on an as-needed basis, such as engineering, purchasing, or stores. These teams will provide answers to problems that some companies have tried for years to solve independently. This team effort is one of the true indicators of a successful TPM program.

OVERALL EQUIPMENT EFFECTIVENESS

The overall equipment effectiveness (OEE) is the benchmark used for TPM programs. The OEE benchmark is established by measuring equipment performance.

Measuring equipment effectiveness must go beyond just the availability or machine uptime. It must factor in all issues related to equipment performance. The formula for equipment effectiveness must look at the availability, the rate of performance, and the quality rate. This allows all departments to be involved in determining equipment effectiveness. The formula could be expressed as:

$$\text{Availability} \times \text{Performance rate} \times \text{Quality rate} = \text{Equipment effectiveness}$$

The availability is the required availability minus the downtime, divided by the required availability. Expressed as a formula this would be:

$$\frac{\text{Required availability} - \text{downtime}}{\text{Required availability}} = \text{Availability}$$

The required availability is the time production is to operate the equipment, minus the miscellaneous planned downtime, such as breaks, scheduled lapses, meetings, etc. The downtime is the actual time the equipment is down for repairs or changeover. This is also sometimes called breakdown downtime. The calculation gives the true availability of the equipment. It is this number that should be used in the effectiveness formula. The goal for most Japanese companies is a number greater than 90 percent.

The performance rate is the ideal or design cycle time to produce the product multiplied by the output and divided by the operating time. This will give a performance rate percentage. The formula is:

$$\frac{\text{Design cycle time} \times \text{output}}{\text{Operating time}} = \text{Performance rate}$$

The design cycle time (or production output) will be in a unit of production, such as parts per hour. The output will be the total output for the given time period. The operating time will be the availability value from the previous formula. The result will be a percentage of performance. This formula is useful for spotting capacity reduction breakdowns. The goal for most Japanese companies is a number greater than 95 percent.

The quality rate is the production input into the process or equipment minus the volume or number of quality defects divided by the production input. The formula is:

$$\frac{\text{Production input - quality defects}}{\text{Production input}} = \text{Quality rate}$$

The product input is the unit of product being fed into the process or production cycle. The quality defects are the amount of product that is below quality standards (not rejected; there is a difference) after the process or production cycle is finished. The formula is useful in spotting production quality problems, even when the poor quality product is accepted by the customer. The goal for Japanese companies is a number higher than 99 percent.

Combining the total for the Japanese goals, it is seen that:

$$90\% \times 95\% \times 99\% = 85\%$$

To be able to compete for the national TPM prize in Japan, the equipment effectiveness must be greater than 85 percent. Unfortunately, the equipment effectiveness in most U.S. companies barely breaks 50 percent: little wonder that there is so much room for improvement in typical equipment maintenance management programs.

A Sample OEE Calculation

A plastic injection molding plant had a press with the following stats:
- The press was scheduled to operate 15 8-hour shifts per week.
- This gave a total possibility of 7200 minutes of run time per week.
- Planned downtime for breaks, lunches, and meetings totaled 250 minutes.
- The press was down for 500 minutes for maintenance for the week.
- The changeover time was 4140 minutes for the week.
- The total output for the operating time was 15,906 pieces.
- The design cycle time was 9.2 pieces per minute.
- There were 558 rejected pieces for the week.

What is the OEE for the press for the above week?

Overall Equipment Effectiveness

1. Gross time available — 7,200 minutes
    (8 × 60 = 480 minutes) × 15 turns
2. Planned downtime — 250 minutes
    (for PM, lunch, breaks)
3. Net available run time — 6,950 minutes
    (1 − 2)
4. Downtime losses — 4,640 minutes
    (breakdowns, setups, adjustments)
5. Actual operating time — 2,310 minutes
    (3 − 4)
6. Equipment availability — 33%
    (5 / 3 × 100)
7. Total output for operating time — 15,906 pieces
    (pieces, tons)
8. Design cycle time — 0.109 minutes / piece
9. Operational efficiency — 75%
    (8 × 7 / 5 × 100)
10. Rejects during shift — 558 pieces
11. Rate of product quality — 96.8%
    (7 − 10 / 7 × 100)
12. OEE — 23.96%
    (6 × 9 × 11)

Figure 1-4. A 12-point data collection process facilitates analysis of OEE information.

A form to collect and analyze OEE information is pictured in Figure 1-4. The equipment availability is calculated in the first section of the form.

The gross time available for the press is entered in line 1. The planned downtime, which involves activities that management sets a priority on and cannot be eliminated, is entered in line 2 (the 250 minutes for the week). The net available time for operation is entered in line 3 (this is actually line 1 minus line 2). The downtime losses, which are all unplanned delays, are entered in line 4. This would include maintenance delays, changeovers (which can be minimized), setups, adjustments, etc. The actual time the press operated is entered on line 5 (this is the difference between lines 3 and 4). The equipment availability (line 6) is line 5 divided by line 3 times 100 percent.

Overall Equipment Effectiveness	
1. Gross time available (8 × 60 = 480 minutes) × 15 turns	7,200 minutes
2. Planned downtime (for PM, lunch, breaks)	250 minutes
3. Net available run time (1 − 2)	6,950 minutes
4. Downtime losses (breakdowns, setups, adjustments)	695 minutes
5. Actual operating time (3 − 4)	6,255 minutes
6. Equipment availability (5 / 3 × 100)	90%
7. Total output for operating time (pieces, tons)	54,516 pieces
8. Design cycle time	0.109 minutes / piece
9. Operational efficiency (8 × 7 / 5 × 100)	95%
10. Rejects during shift	545 pieces
11. Rate of product quality (7 − 10 / 7 × 100)	99%
12. OEE (6 × 9 × 11)	85%

Figure 1-5. Overlaying world-class standards on the baseline data of Figure 1-4 shows a productivity improvement of 350%.

The equipment operating efficiency is calculated in the next section. The total output for the operating time is entered in line 7. The actual design cycle time (this number must be very accurate) is entered on line 8. The operational efficiency is calculated and entered on line 9. The operational efficiency is line 7 (the total output) times line 8 (design cycle time) divided by line 5 (the actual operating time) times 100 percent. This number should be evaluated carefully to ensure that the correct design capacity was used. If the percentage is high or exceeds 100 percent, then the wrong design capacity was probably used.

The quality rate is determined by the total output for the operating time (line 7) minus the number of rejects for the measured period (line 10) divided by the total output (line 7) times 100 percent.

In the sample, the availability is 33 percent, the operational efficiency is 75 percent, and the quality rate is 96.8 percent. The OEE for the press for the week is 23.96 percent.

What do these conditions mean? What do the indicators show the typical manufacturer? The answers are evident when a second model using the same press is examined. In Figure 1-5, all the parameters are set at world-class standards to give an OEE of 85 percent. As can be quickly observed, the major improvement is in the total output for the operating time (line 7).

The press now will make 54,516 parts, compared to 15,348 with the 23.96 percent OEE. Since the resources to make the parts (labor and press time) are the same, it makes the company more product and ultimately more profits. With the press operating at an OEE of 85 percent, the same productivity results as if 3.5 presses were running at the 23.96 percent OEE. The potential for increased profitability and ultimate competitiveness is staggering.

Total productive maintenance can have a positive impact on any company's productivity and profitability, as long as the entire organization is willing to change its culture and the way in which day-to-day business is conducted.

THE TIMKEN EXPERIENCE

At The Timken Company's Gaffney Bearing Plant, implementation of TPM has yielded benefits both measurable and intangible. The company is one of the pioneering firms in America to introduce TPM to their operations to counter rising costs and eroding profitability. Timken recognized they had a valuable resource in their work force and that if they could tap into the talents of their people on a day-to-day basis they could, in fact, improve their cost base, increase profitability by boosting uptime of their equipment, and enhance the quality of products coming off that equipment.

Timken went to work on these goals, introducing TPM as the backbone of an ambitious program they call Vision 2000. It is a plan and a vision of where the company wants to be by the year 2000—the best performing manufacturing company in the world, not only in bearings, but in general manufacturing. The company's goal is backed by a management commitment to provide the best, most advanced tools available for training their people, improving processes, and fostering team concepts through work force empowerment corporatewide.

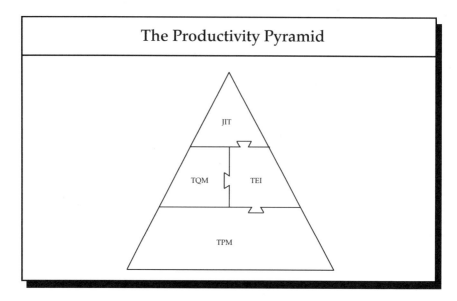

Figure 1-6. TPM at Timken is the foundation of world-class manufacturing.

Vision 2000 is a program consisting of four key strategies:
- Total productive maintenance,
- Total associate (employee) involvement,
- Total quality management,
- Just-in-time manufacturing.

Timken believes that these are the major strategies, the major building blocks that will lead it to achieving the mission of becoming the world's best performing manufacturing company by the year 2000. TPM, in Timken's view, has to be at the foundation of that structure, the pyramid of world-class manufacturing. The interaction of the human resource and the machines is at the essence of Timken's manufacturing process. The link is an equipment management strategy that connects people with their machines in a way that enables them to see they are making a difference, that they are having a positive impact on the outcome of the enterprise. That belief is what moved Timken to settle on TPM as the first and primary building block of a world-class manufacturing pyramid, as shown in Figure 1-6. TPM gets to the core of what actually creates value for customers.

Thus far, TPM has helped the company reduce overall cost in several measurable ways. With 40 percent of the equipment in their "green" machining operations (their pilot area) covered by TPM, they have recorded a drop in lubricant cost from around $12,000 a month to about $5,000. Their estimated cost savings when all green machining is covered by TPM is from $100,000 to $140,000 per year in one department alone.

CHAPTER 2

Elements of Successful TPM

> *Within an organization, there may be a number of different plants with different cultures that have different levels of willingness to accept change. After the decision has been made to implement TPM, a clear understanding of the current culture and its receptiveness must be established.*
>
> —Perry D. Stephens, Chief Manufacturing Engineer,
> The Timken Company, Gaffney Bearing Plant

The first hurdle to overcome before pitching TPM to upper management is taking a close look at where you are now in terms of corporate culture and willingness to change. Once this has been assessed and the program's starting point set, the next hurdle is selling upper management on TPM's long-term positive effect on the overall bottom line. It will take an environment where not only do you have the technical expertise, but also a climate in which people are excited enough to become involved and want to make a contribution. Most of the ongoing improvement activities are highly dependent on employee involvement and employees taking ownership of equipment and processes. TPM fits quite nicely into that scenario.

Employee empowerment and involvement are absolutely essential to TPM, and it will take top management commitment, an adequate budget, and changes in corporate culture to make it happen. Unless workers are given the power to act on problems, unless they are given the opportunity to become involved, unless they are given the authority to make things happen, total productive maintenance will be a futile effort at best.

COMMITMENT

The importance of management commitment in a TPM program is that TPM is an empowering process. As such, one of the most difficult things to struggle with on a day-to-day basis is convincing workers that, first, they are empowered to do things that before they weren't and, second, management is serious about TPM.

The problem of empowerment is one of getting the workers to test the water—to convince them that their ideas are important and that they are now decision-makers in the company and that management is there to back them up.

Management commitment can be exhibited in any of a number of ways.
- By being accessible, on the factory floor and in the office.
- By sending TPM teams to national conferences. This sends the message that management is willing to invest in its people—production workers seldom get the opportunity to attend conferences of any kind.
- By staying involved, taking an active interest in what the TPM teams are doing on the plant floor.
- By keeping visibility high. Publishing articles in company newsletters, etc. Recognizing significant TPM achievements. Keeping communication channels fluid and open. Providing the means to have workers' voices heard.
- By demonstrating that management has a team mindset, as opposed to an autocratic one.
- By providing an environment in which management is open to change and willing to permit workers to plan for and implement change.

COST

Like all other programs, TPM comes with a price tag. From the very beginning, it must be impressed on senior management that launching a TPM program will cause an initial increase in costs due to accelerated maintenance activities, team-building training, and technical training. Start-up costs will be incurred in assessing current equipment effectiveness and "baselining" pilot equipment in the plant. Introducing the plan to the entire work force and communicating it on a regular basis will require additional outlays for newsletters, communication centers, etc.

But TPM's long-term payoffs will overwhelm costs. To the extent that downtime of your equipment can be reduced, you are going to save money—you are going to keep production running. To the extent that the performance of your equipment can be enhanced, you are going to maintain throughput, you are going to improve the quality of your product. To the extent that your equipment is adequately maintained, you are going to keep it in service longer and thereby reduce your capital expenditures.

CULTURE

Company culture is one of the most critical aspects determining if TPM will be successful or not. The company that truly believes in using the talents of its people is more likely to have a successful TPM program than one still hanging on to the autocratic principles of "Taylorism." Experience has shown that workers thrive on involvement in an environment where they are treated as productive individuals who have a voice in their workplace.

"Productive" is TPM's middle name. Productivity is fostered where management is willing to provide the latitude for people to try new things, even if they fail occasionally. TPM requires a culture where there is a commitment to change, a commitment to ongoing improvement, and a commitment to treating each individual as a valued employee.

Implementation of TPM will have a profound, positive effect on the culture of a company. It will change the culture. It will change relationships across organizations of the company. It will distribute decision-making and disperse the authority base.

A definite correlation exists between management style and the culture of an organization. How people are led and managed affects how they feel about the company and how much discretionary effort they contribute. It also affects the health of the company.

Conventional practice in recent years has seen many companies restructure and downsize their operations. Those that could not compete successfully are gone. Among those that survived, there is a common denominator: all recognized that they must change, and the change involved the fundamentals of the way they conducted their businesses. In some companies, culture changed dramatically. For most, the new culture evolved. In all, a more participative climate emerged.

Buy-in by everyone in the company is central to creating a climate for TPM. Each person must recognize the need for change and be dedicated to making it happen. The need for change does not necessarily mean that the company is on the verge of going out of business. It does mean, however, that everyone in the organization must realize that changes are necessary to maintain a competitive advantage, to make the company—and themselves—prosper. Status quo must be seen as a sure way to weaken the company.

There is no magic formula for making changes, but starting at the top of the organization works best. Senior management must have a vision that's contagious. Each company must develop its own vision, which must be translated into strategy and tactics. Measurable goals and objectives have to be developed. Buy-in and commitment must be gained from everyone in the organization to achieve the vision and, as time goes on, the vision will need to be adjusted to meet new challenges and opportunities. This will cause further changes. This change continuum will become a way of life, because it has no end.

Indicators of successful change in organizations form around certain common characteristics. Change in this context means the company will likely succeed in implementing a strong total productive maintenance program. Some of the characteristics may not be possible in terms of what is practical, but collectively they form a good starting point for understanding where the organization of a company stands. These common characteristics of successful organizational change include:

1. *Customer Focus.* The priority of everyone in the TPM program must be the internal customer. The maintenance department's customer is the machine operator, who expects his equipment to be serviced and repaired with regularity; the operator's customer is management responsible for throughput rate, who expect equipment to have zero downtime; the manager's customer is the company's customer, who expects zero-defect products quickly and at competitive cost; the final customer is the owner/shareholder, who expects the company to be profitable and have production-ready assets.

2. *Management Commitment.* The bottom line is that management must "walk like they talk." Actions must be directed toward improving overall equipment effectiveness. Management cannot vacillate in this; workers pick up on this and quickly assume that management is not serious.

3. *Change.* Change should be taking place on a wide scale. Not all change works, but people should be—and generally are—willing to try new things.

4. *Management Philosophy.* Old management styles should be disappearing, replaced by more involvement of the workers. Empowered workers feel that they are a vital part of the company.

5. *Risk Taking.* Risk should be recognized as a part of the business climate. People should be able to take risks and know that they will not damage their careers. Because of this, problems will be solved quickly.

6. *Information.* There should be a good flow of information within the company. People should feel informed and trusted. They should have the information needed to do their job and to help in planning the future.

7. *Roles.* The role of each person in the company should be clearly defined. Everyone ought to be aware of where they must go for help or information.

8. *Teamwork.* The organization should foster team spirit. Controls should be relaxed to permit self-direction of tasks and projects by people working cooperatively.

9. *Strategy.* The strategy of the company should be clearly represented in the way resources are intermeshed. Carefully planned integration of technology, organization, and people makes a strong message for the importance of each individual in the organization.
10. *Tasks.* The form of the organization should be flexible enough to do various routine tasks in an effective manner.
11. *Decision-making.* The organization should be designed to drive decision-making to the lowest level possible. The best decisions are usually made by those who will be personally affected. Attention should be placed on an organization that can make decisions quickly.
12. *Stability.* To encourage a feeling of belonging and dedication, the organization should not be changed frequently without good reason. Where a change is required, extensive efforts must be made to accommodate the change and to communicate to all the rationale for the change.
13. *Innovation.* The organization should provide for the constant development of innovative approaches to improve, enhance, and strengthen the TPM process. Much of the grist for this development will come from the shop floor. Let it be heard and recognized.
14. *Trust.* The organization should promote a high degree of trust among its employees. One part of the organization must not be pitted against another in an adversarial relationship. Teamwork and cooperation must prevail throughout the organization.
15. *Problem-solving.* The company should have a problem-solving process that is widely understood and used.

The common thread binding these characteristics of successful change is the individual worker as the focal point in a team-driven organization. By utilizing people's talents and ideas, not just their physical abilities, a great deal of positive change can be effected.

Those involved with the equipment on a daily basis are the primary equipment stewards, or caretakers, in a TPM culture. The most receptive culture for TPM implementation is one where people at all levels understand the business environment in which they function, why they are there, the mission of the company, and what kind and level of competition they are facing or expecting to face. And if the workers are prepared to make the changes necessary in terms of their work habits to ensure the long-term survival of their organization, an ideal TPM culture is defined. Once into TPM the ideal culture is one in which those closest to the equipment have assumed the basis of "ownership" of their equipment, much as they own their cars, respecting it and caring for it like they would their own property. Above all, to maximize TPM's benefits, employees *must* be empowered to manage their own areas. Operator/maintenance teams are the key. Without them, you don't have TPM.

Operators have the most knowledge about how a machine or process works. They know what to do to increase the profitability of the company at the shop-floor level, to make the company competitive worldwide. That's why it is absolutely essential that shop-floor workers be involved in the decision-making process, that they have the facts and information at hand to make informed choices.

Armed with proper and sufficient information, workers don't have to wait to get something done. They don't have to wait for the process of going up the ladder and then back down. They go crossways—across functions—saving a lot of time. Efficiency is the result.

But even as the TPM wave gains momentum, the core dynamic is management's continued vigorous support and recognition of employees' efforts.

CHAPTER 3

Getting TPM Going

The first step [toward TPM] is to make sure your employees are willing to get involved. If you have a lot of worker involvement you will have a much easier time implementing TPM. Working together is going to produce a lot better result than working separately.

—Wayne Vaughn, Facilities Manager, Harley-Davidson Company

The driving force of TPM is its goals, as shown in Figure 3-1. Goals get TPM going, sustain it, and keep it going. The primary goal—to ensure that equipment performs at full capacity—is advanced via a maintenance program designed for the life of the equipment. This requires support from all departments involved in the equipment's use, input from people at all levels of the company, and small group activities focused on continuously improving the process, the equipment, and the throughput.

One of the first things a company needs to do to get TPM going is to communicate very clearly the results it wants to achieve through the TPM effort. A common mistake is to say "we want to do TPM for TPM's sake,"

TPM Goals

- Ensured capacity
- Lifetime maintenance program
- Departmental support
- Employee input
- Small group activities
- Continuous improvement

Figure 3-1. Establishing and communicating goals is a crucial step in TPM implementation.

that we're doing it because the Japanese are doing it. "Me-too-ism" doesn't work. Commitment and communication does.

TPM educator Robert Williamson cites the importance of doing your homework. His counsel: read everything you can get your hands on about TPM as practiced in Japan. By so doing, you will find out that the elements of TPM aren't really new. It is planning, scheduling, preventive maintenance, predictive maintenance, training, operators learning to run their machines effectively, pulling detailed inspections of the equipment, conducting small group activities, and common sense types of things companies are doing now. So in preparing the organization for TPM, learn about TPM and show what you are already doing as part of your standard operation.

SOWING THE SEEDS

Generally, the TPM process is initiated by a "champion" or group of champions who push the program and educate key people within the organization to its philosophy, the improvement it brings to day-to-day operations, and the potential savings and increased capacity it can generate. That's the high end. Before that can happen, some fundamental spade work must be done: before TPM can be pitched to upper management, the company's underlying culture and workers' willingness to change must be honestly evaluated. This is essential to determining the program's appropriate starting point.

Once this has been done, and the program's starting point set, the next hurdle is actually selling upper management on TPM's long-term positive effect on the bottom line. It cannot be overstated that for TPM to be successful, the full support and *visible* total commitment of upper management is absolutely essential. An effective strategy is to convince key individuals at various levels of the organization to become strong supporters and help advance the cause. A strong base makes a strong case.

Williamson suggests focusing on the middle tier of management. Middle managers, especially those in first-line supervision and the next level, are in a critical position to support TPM. They are in a position, actually, to make or break the effectiveness of TPM. More people report to first-line supervisors in a company than any other level. Thus, the first-line supervisor is really in charge of most of the ingredients of productivity: the people, the assignment of schedules, the assignment of work, directing people, coaching, counseling, seeing that they get the right education and training. The support that this level of management gets from the next level is most critical. There has to be a clear understanding of the new roles in a TPM environment. It is best to involve middle management and first-line supervision in the definition of your company's TPM program, in the early understanding stages before you roll it out to the plant floor. First-line supervisors must be committed to achieving the results of TPM, and they must have the opportunity to share their ideas for improving equipment effectiveness in the design of the company's TPM strategy.

As stated in chapter 1, top management must have a vision for TPM. But making the vision come to be is not easy. It is much easier to talk about change than to make it happen. Shared visions are difficult in view of company politics and empire-building; turf battles between activities still and will occur. Yet, for TPM to be realized, all activities in a company must occur in unison. Overcoming these barriers and others is a challenge for top management. Without the active involvement of senior management—especially at the outset—removing the barriers will be an immense task. There have been efforts to change from the bottom up in an organization, but without a strategy for dealing with change at the top, the risk of failure is high.

CULTURE SHOCK

The type of change called for in TPM is especially difficult because in many respects it pervades the fundamental nature of the company's work culture. It reaches through and affects the entire organization. Arguably the most difficult thing to change is the culture of the organization—the way it does things and how people interact. Quite likely, premises and the principles on which the company is organized must be revised. Activities must adapt to what is required to achieve the vision of what the company wants to be. All this must be supported and orchestrated by the top person to help smooth the transition. In a sense, this likely will take more stamina and commitment than companies or individuals have been asked to muster ever before. That's why it is imperative that the required level of energy and dedication be made unmistakably clear to the front office.

When your company makes the decision to undertake a TPM project, several things have to happen to prepare the workplace for it. Certainly the group that is going to be instrumental in rolling out the program has to become quite knowledgeable of the process and methodology of TPM. Once that happens, supervision must be tutored and brought aboard.

Questions of the uninitiated—from senior management to line worker—have to be answered. Why use TPM? What is its intrinsic value? What can be gained from it? Is it worth the effort? What does it do for me?

THE STARTING LINE

A flurry of "kickoff" activities will inaugurate the program. A company-wide announcement of the decision to implement TPM should be made so that everyone in the organization is made aware of TPM's benefits. A pilot area should then be chosen from which TPM will build, and a team formed within that area. Team building may start at this point, depending on your company's level of involvement with other world-class activities.

This is when well established and effective communication channels show their merit. Often, companies have tried improvement efforts, but fail to communicate up front and find their efforts fruitless from the start. Conversely, successful companies have communicated effectively with their people on why given programs are being implemented, what the desired results are, what the competition is doing, and that "we need to do it to meet our customer expectations by improving our equipment effectiveness." This type of communication is more likely to foster interest and buy-in.

But communication doesn't end there. Williamson warns to be prepared for some predictable questions from workers. "Why change? We've always done it the other way." At this point you need a clear reason for change and an effective plan for communicating it. The answer should include how it will affect individuals' jobs.

Fundamental to human nature is anxiety about personal and job security. Such anxiety must be put at ease at the beginning, Williamson claims. A common question is "Will some of us lose our jobs or have to take pay cuts because of TPM?" The answer is obviously no, but crafting the communication to impart that message to those most concerned is the real task. Workers know of the many programs and processes in place in companies that are designed specifically to reduce head count and control costs. TPM is not one of them and should be communicated as such; its goal is improving equipment effectiveness, and to achieve that, all people working together are needed.

Then there are those who will ask, "Do I have to participate? I'm getting close to retirement and I don't want to start doing a lot of new things." To those, the answer is no, you don't have to take part, but if you are in a TPM pilot area, you'll have to move to an area in the company not currently in TPM mode. Everyone in the TPM area must be committed to making it work, and that mandates buy-in and participation.

THE ELEMENTS OF TPM

At the very foundation of TPM is partnership. TPM focuses on partnership among the manufacturing people, maintenance, engineering and technical services to improve the overall equipment effectiveness.

Baselining

The development of a TPM plan itself begins with recording the current status of your organization. This evaluation will include more than just the maintenance organization; it should embrace all parts of the organization involved in the operation, maintenance, design, and purchasing of the assets of the company. The most important benchmark for TPM is the OEE. By establishing (baselining) the current OEE for the plant assets (as represented initially by the pilot equipment identified), the organization is given a goal to work toward. The primary objective is to optimize the equipment effectiveness. Any other objective fails to unite the various parts of the organization into a competitive, focused unit.

Baselining provides the starting point from which to calculate improvement. As such, it also provides the means to demonstrate to your management that you are making progress, that you are saving money, that productivity is affected in a way that improves the bottom line. Baselining also shows irrefutably to the production/maintenance teams that they are accomplishing something. This keeps interest and enthusiasm levels high.

Baselining involves documenting as much current data on equipment as possible. From the current, or baseline, status you can show management where the equipment was in terms of productivity, where it is, and where the trend patterns are pointing. Such data could include quality information of the equipment, time data, maintenance downtime histories, setup times, and procedures. Other documentation might include photos showing the physical condition of the equipment at program start to compare with similar photos after TPM implementation.

Piloting

Selecting a pilot area is an important step in the TPM sequence. Rather than attempting to implement TPM plantwide right from the start, it is good counsel to start small and target pieces of equipment identified as critical to a production process.

Preventive Maintenance

When considering the maintenance component of the organization, refer to the evolution chart (Figure 1-3, page 7) for guidance in determining the basic foundation. The first focus is on a preventive maintenance program. Without a good preventive maintenance program, the equipment is never maintained at a level sufficient to ensure that the organization has assets capable of producing a world-class product or service.

Predictive Maintenance

Implementing a predictive maintenance program as the next step establishes a formal method for monitoring equipment and ensuring that wear trends are documented. In this way, equipment can be overhauled, with worn components changed, before a failure occurs. Figures 3-2 through 3-6 show typical inspection forms for hydraulic systems, pneumatic systems, belts, chains, and general equipment, respectively.

Hydraulic Inspection	Item	OK	Needs Repair
Hydraulic pump	Proper oil flow?		
	Proper pressure?		
	Excessive noise?		
	Vibration?		
	Proper mounting?		
	Excessive heat?		
Intake filter	Clean?		
	Free oil flow?		
Directional control valves	Easy movement?		
	Proper oil flow?		
Relief valve	Proper pressure?		
	Excessive heat?		
Lines	Properly mounted?		
	Oil leaks?		
	Loose fittings?		
	Damaged piping?		

Figure 3-2. A typical inspection form for hydraulic systems.

Pneumatic Inspection	Item	OK	Needs Repair
Compressor	Proper air flow?		
	Proper pressure?		
	Excessive noise?		
	Vibration?		
	Proper lubrication?		
	Excessive heat?		
Inlet filter	Clean?		
	Free air flow?		
Directional control valves	Easy movement?		
	Proper air flow?		
Muffler	Proper air flow?		
	Proper noise reduction?		
Lines	Properly mounted?		
	Air leaks?		
	Loose fittings?		
	Damaged piping?		

Figure 3-3. A typical inspection form for pneumatic systems.

Belt Inspection	Item	OK	Needs Repair
Belt sheave	Sidewall wear?		
	Dirt in sheave?		
	Alignment		
	Mounting		
	Support bearings		
	Guards		
Belt	Stretch?		
	Narrow spots?		
	Tension		
	Tracking of belt		
	Contamination		

Figure 3-4. A typical inspection form for belts.

Chain Inspection	Item	OK	Needs Repair
Chain sprocket	Hooked teeth?		
	Wear on tooth sides		
	Alignment		
	Mounting		
	Support bearings		
	Guards		
Chain	Elongation		
	Side plate wear		
	Tension		
	Engagement of sprocket		
	Proper lubrication		

Figure 3-5. A typical inspection form for chains.

Equipment Inspection		Item	OK	Needs Repair
Main equipment cleaning	Contamination	Dust		
		Dirt		
		Oil		
		Foreign material		
Main equipment wear	Mechanical	Loose nuts		
		Missing nuts		
		Excessive play		
Main equipment lubrication	Oil	Proper level		
		Leakage?		
		Covers?		
Main equipment lubrication	Grease	Fittings clean?		
		Proper grease?		
		Proper amount?		
5 senses inspection	Temperature?			
	Noise?			
	Vibration?			
	Visual?			
	Odors?			

Figure 3-6. A typical inspection form for general equipment.

Computerized Maintenance Management System

The subsequent improvement in the organizations's ability to service the equipment is followed by an increase in the amount of data used to perform failure and engineering analysis. This highlights the need for a computerized database for tracking and trending equipment histories, planned work, the preventive maintenance program, the maintenance of spare parts, the training and skill levels of the maintenance workers, etc. The systems commonly used for this are called computerized maintenance management systems (CMMS), a comprehensive relational database accessible to the entire organization.

Continuous Improvement

As your TPM program matures, the company will see how much support maintenance provides other advanced programs. JIT programs are enhanced, since now the equipment runs when it is scheduled and produces at the rate it was designed to operate. Total quality management (TQM) programs are enhanced, since the equipment is stable and produces a quality product every time it operates. Total employee involvement (TEI) programs are matured, since most of the problems the employees are currently having relate to the equipment. This makes their responsibilities easier to meet and leads to more motivated workers concerned with maximizing their competitive strengths.

HOW HARLEY-DAVIDSON DOES IT

Harley-Davidson, at the outset of its TPM program some years ago, contracted with a maintenance consultant to help them identify a pilot area and establish a lubrication program for equipment in that area of its plant. The company's focus was to reduce equipment downtime caused by lubrication problems. The pilot area established the baseline from which they would articulate their goals and establish standards.

One of the first steps the TPM team took was to conduct an analysis of how much additional capacity the company could get from their machinery if they improved OEE from where it was at baseline to the 85-percent goal that in TPM circles is commonly accepted as optimum. That goal was fully in line with overall corporate investment objectives which were 35 percent return on investment. In fact, the 85-percent TPM goal actually exceeded corporate goals.

Today, the company is seeing those kinds of results. Because other programs to improve capacity also contribute to the improved corporate performance, it is difficult to isolate TPM and put a discrete percentage figure on its specific contribution, but the company is convinced of TPM's benefits.

In conjunction with its TPM program, Harley-Davidson has in place its version of just-in-time, which they call their "material as needed" (MAN) program. MAN teams work on implementing improvements in the demand system, coordinating efforts with the TPM team. The MAN team conducts a project in a new pilot area, and the TPM team follows close behind to implement the maintenance program.

The programs implemented at Harley-Davidson have produced intangible benefits as well. Tradespeople have been freed up to do more predictive maintenance, since reactive maintenance is no longer an issue. Breakdowns are at a minimum, which allows more time for predictive maintenance, which cuts the potential for future breakdowns even further.

As the Harley-Davidson case attests, starting TPM small in a pilot area increases its chance of success on two levels: first, it gives the team time to grasp and internalize TPM's priorities and goals, and, second, targeting critical areas increases the likelihood of achieving improvements quickly.

The catalyst for success, however, is teams. Within the pilot area, all people directly involved in the equipment's operation—operators, maintenance personnel, supervisors, manufacturing engineers, and quality people—need to be organized into teams and trained in TPM methodology.

CHAPTER 4

Building TPM Teams

It is fantastic to see people working together that normally would have never worked together before. Working as a team, solving problems, addressing problems, making modifications. It excites me to see that, because there is nothing but positive results.

—Curtis J. Hapner, Facilitator, Equipment Services,
Magnavox Electronic Systems Company

TEAMWORK IN TPM

There is no better way to effectively involve an organization than to create natural teams of people who work together toward achieving a common goal developed by them for themselves. The legitimacy and validity of the team concept derives from its inherent synergy. Teamwork allows people to draw on their mutual strengths to reach a successful conclusion that on an individual level would not be attainable. Although it is a natural part of the business process, it is a learned process. Teamwork requires the development of trust throughout the organization and requires that team members become interdependent, with a willingness to cooperate toward achieving common goals. Such a transformation requires structure, training, and coaching of the organization and the teams.

Teamwork—the formation and use of teams—is not just a process. It must become a part of the business culture. In a TPM culture it is essential. The use of teamwork in business is a new paradigm and will require a change in the business culture.

A change in the business structure to support teams is a critical part of the social system or cultural change in the business. The new structure must be based on a clearly understood TPM business vision, mission, and plan and an understanding of that plan by everyone in your company. In addition to the work force understanding the plan, senior management must become versed in the process of teamwork. This understanding should include firsthand experience as a team member. Appropriately, then, teamwork needs to start at the top of the organization and cascade down, rather than begin at the bottom and swell up. Where a labor union exists, labor leadership must be involved in the process at the outset as if it were a function or agency of your company, which, in reality, it is. Failing to meet these prerequisites was a principal cause for the failure of many quality circle efforts in the U.S. in the 1970s and 1980s. Without a top-down culture change and support structure—and without labor's buy-in—TPM teams cannot work.

THE CRITICALITY OF CHANGE

Changing the business culture is the most difficult challenge you'll face in creating a total productive maintenance environment. Culture resists change because culture represents the sum of all of everyone's habits. It is difficult to change just one habit, much less change all. Language is a type of habit. We learn it before age 10 and feel quite comfortable with it as adults and generally don't think much about it when we use it. But what if we moved to Mexico or Japan? We'd be facing a major habit change.

So it is with TPM. Team members will have to change individual habits. They will have to change communication patterns, their need for control of outcomes, their willingness to allow others to do work and make decisions. A sharing of information will have to occur among production, maintenance, and engineering where before the information was considered confidential, or at least proprietary.

Trust

Critical to both TPM and the teams that will make it work is trust. If the organization has not developed a culture in which there is trust sufficient for decisions to be made "on the floor," that trust must be developed. If operators have never before been empowered to manage processes, trust in their skill and judgment must be assured and made known to them. The formation of teams and initiation of education in business values as well as in the mission, objectives, and goals of TPM will contribute to raising the level of trustworthiness of operators. During team development, operators will learn the meaning of *empowerment* and *ownership* and be trained in the technical skills needed to maintain their equipment. They must fully comprehend their roles and their expected contributions to the TPM effort and where those roles and contributions fit within the company-wide TPM program.

Bonding as a Team

The process of team building is, in reality, that of laying the cornerstone of a TPM effort. It is a process in which several diverse individuals are bonded in a group toward a common goal or set of goals. The strength of the bond determines the effectiveness of the team: the goals, its direction. Gaining agreement on the common goals of the TPM effort is an important element in the bonding process. And, interestingly, it is the differences of each member—the individual skills, knowledge, ideas, and decision-making capability—that helps assure well-tested solutions.

Members of your TPM team will come from different functional areas of the company. Some members may even be from outside the company—equipment suppliers or consultants, for instance. Each member brings to the group needed skills that together with other members' contributions combine to reach the goal.

Many organizations pursuing TPM use a facilitator for team building. It works. Approaches differ, of course, depending upon the facilitator, but they produce the same result—bonding of the team members.

The Necessity of Synergy

In the business world and, specifically, in manufacturing, the move from an autocratic type of directed work force to a more team-oriented universe is as much a product of necessity as it is the creation of organizational development science. The radically changing nature of markets has dictated it. Intense and incessant global competition daily challenges the way we do things, and change has become the manufacturing manager's only reliable constant.

Teams have emerged as one of the most effective changes. The realization that two heads are better than one and three are better than two limits risk in decision-making as team members examine all possible outcomes of decisions and their implications. The organization benefits if it can exploit the synergy implicit in group intellect around the common goals and objectives that are aligned with the goals of the organization as a whole.

TPM teams are formed to solve problems. Since no two problems are alike, teams must have the flexibility to tap different skill sets and call in specialized personnel when necessary. This dynamic interaction between all levels of the organization fosters open communication and creates an awareness of the value all employees add to the manufacturing process.

The process of becoming an effective TPM team is developmental. This aspect of teamwork manifests itself in many everyday examples. It is an accepted norm for baseball and football teams to undergo a long practice period prior to playing games that count toward the season record. Too often, businesses will bring a group of individuals together as a team and allow little or no opportunity for the team to develop before being "thrown

to the wolves" and expected to perform efficiently. When brought together for the first time, teams must accomplish several tasks before undertaking the pursuit of their objectives. They must:
- Reach a mutual understanding of what the mission of the team is.
- Decide the roles of the various members of the team.
- Decide the methods by which the team will operate.

Roles of Team Members

From a generic perspective, roles on a manufacturing team are defined by assignment and team consensus. There are four basic roles of the various members on a business team:
1. Team leader.
2. Team facilitator (coach).
3. Team recorder.
4. Team members.

In most cases, teams will have additional assigned tasks, but each team should define these roles and select or assign people to fill them.

Team Leader. The team leader focuses on teamwork content. He or she will help maintain focus on the TPM team mission and has the primary responsibility to act as liaison between the team and management (who, after program implementation, back away from the day-to-day activities of the program). As liaison, the team leader keeps management informed about how the team is performing, based on the team mission or goal.

Likewise, the team leader will usually be the conduit for communicating business news to the team. He or she arranges for the team's needs from the business and informs management of the team's progress. In addition, the team leader is an active member of the team and should be elected by the team, not appointed.

Team Facilitator. The facilitator focuses on the processes of the team. In this role, this "coach" is an outsider to the team and should remain neutral. In the TPM context, this person could well be one of the company's TPM "champions." The primary task of the facilitator is to help the team operate smoothly. The facilitator is the key person to prepare for meetings, typically planning logistics for meetings, creating a tentative agenda and plan for completing meeting work, and informing team members of times.

During meetings, the facilitator helps the team keep on track by initiating discussions and making sure they stay focused and on schedule. The facilitator's focus should be on the process of the meeting, not necessarily on the content of the discussions. When the team is ready for introduction of a new tool or skill of problem solving or process improvement, training in these tools or skills is the prime responsibility of the facilitator.

Team Recorder. The team recorder is not only the team historian, but also the key to keeping track of the team's performance. Sometimes it works well to divide these duties among several people. The recorder maintains records of the meetings as they happen, including action plans and commitments. Besides the meeting minutes and problems and action lists, the recorder should maintain team records, meeting agendas, research data, and other appropriate information such as mission measurement data. These records and information help in tracking the team's progress.

Team Member. Team members, other than the leader and recorder, also have significant roles in accomplishing the TPM mission. It is essential that all team members *buy in* to the team mission and share in the ownership of TPM. If any team members have goals that are not in harmony with those of the group or if they have overriding accountability to others outside the team, they may have priority conflicts that can affect their contribution to the TPM team goals and reduce the effectiveness of the team. As partners, all team members need to be involved and take responsibility for achieving the TPM goals. Being involved will include volunteering for assignments and contributing knowledge to the team.

Less formal, but equally as effective, are flexible teams. At Magnavox Electronic Systems Company, where most manufacturing areas are broken down into cells, anyone directly involved with the operation of cells is part of a team. TPM teams are formed from members of the cells, which include all of the operators in a given cell. In addition, a maintenance person is assigned from what Magnavox calls its Equipment Services Department. Supervisors are also team members as are manufacturing engineers and quality specialists when they are involved in the operation. Moreover, any team member has the right, after the team is formed, to call in whomever they think they need to solve problems. This type of flexibility often results in team makeup changing from week to week, month to month.

Commonality in Diversity

There is no such thing as a "canned" team, a template team. Each team within a company is going to be different, to be unique. Each team will find its level of commonality and its level of harmony. Each member will have different ways of approaching things—solving problems, making decisions, arriving at conclusions. What's important is to focus on commonality by noting such things as how frequently words like *we, our,* and *mine* are used.

Teams Take Time

In new team environments, it takes time for ideas to percolate up and anxiety to settle out. The initiation of teams and the development of the needed support structure is a major challenge for the business organization just beginning the move up to total productive maintenance. Change of this magnitude does not happen in six months, and in most cases not even in a year. The cultural changes will take varying times, but in a traditional organization normally one to two years is a reasonable estimate for developing the communications networks and elevating trust levels to a point where the culture is receptive to the responsibilities of empowerment. It will take three to five years for the natural variations in organizational readiness to level out and the TPM process to be considered a business culture.

CHAPTER 5

Learning the Ropes and Staying Abreast

Training is the catalyst that supports the cultural impact of what operators and maintenance staff are going to be asked to do so that TPM becomes their culture. On-going training is essential to keeping the system evolving and viable. New technologies, whether they be in trend analysis, management leadership styles, or production flow will demand that you keep abreast of changes.

—Ed Stanek, President, LAI Maintenance Systems, Inc.

Training makes TPM work. Companies that have been through TPM implementation, that have achieved significant results, affirm that training is one of the key elements of TPM. The entire company needs to understand new ways of thinking about work, new ways of doing work, new ways of working together, new ways of improving equipment effectiveness. They need to understand how they can do some things that in the past they weren't able to do. Training is the enabler, and critical to success is to train, train, and train again.

Attaining TPM's premier goal—to optimize equipment effectiveness—requires educating the entire organization to TPM philosophy, and training production, maintenance, and engineering to implement new methods of conducting day-to-day business. Yet training is a company-specific function and can take many forms. It might be a comprehensive team-driven effort taking anywhere from 6 to 24 months or a course of multi-day training sessions during the program's inaugural phase. Ideally, it would be a concentrated effort at the outset, followed by refresher and new-employee orientation training on a continuing basis. The important point is not so much the length and intensity of the training—different companies have differing needs and various levels of training already in place—but commitment and perseverance. Strategic decision-makers in your company must understand that with their commitment to TPM, workers will have to be away from their production posts so that they can learn the skills they need to increase productivity.

BEGIN WITH THE BASICS

Training for TPM comes in many flavors at several levels. When embarking on TPM, your organization must evaluate its culture to determine if it is receptive to TPM. If it isn't, basic personal skill development sessions—team building, problem solving, or conflict resolution—may be needed. Teams, composed of members from several diverse functions, now expected to perform more effectively in a harmonious manner, may need specialized team training before they even get into TPM activities, especially if the team concept is new to your company. Given that the company must change its culture to foster total productive maintenance and that resistance to such change will be real and evident, massive training is needed in the skills necessary for the new ways of doing business. But before jumping from the boardroom to the classroom, you need to develop a training approach and draft a training strategy to address *all* the needs of TPM as they relate to overall company strategy.

Approach to Training and Education

The approaches for identifying training and education activities today are different from those used by most companies in the past. For the most part, preferences about training formerly were provided by the individuals or departments who desired the training. Management determined the specific training to be delivered, based largely on those preferences.

Moreover, training in the past was decided through a highly decentralized decision-making process at the lower levels of an organization. There was little consideration of widespread training in topics that would be best for the company. This type of fractionated training was directed mostly toward what was good for the individual and the function in which the person worked.

But the focus of training has changed in recent years, prompted largely by unprecedented competitive pressures. The very survival of many companies required a change in the way they did things. Fundamental change was needed to close the gap on quality and costs, and massive, coordinated training efforts were initiated that involved all employees.

A third dimension has been added to the training paradigm: training now encompasses those elements needed by the company to meet its strategic goals. Meeting those goals is absolutely necessary to assure survival in a highly competitive environment. In reality, the third dimension does not conflict with the other two; to survive, a company today also needs to provide training in what is best for an individual and what is best for a functional area. The three dimensions of training are illustrated in Figure 5-1.

Learning the Ropes and Staying Abreast

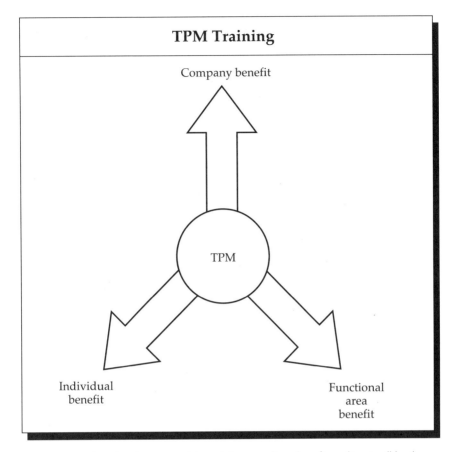

Figure 5-1. With TPM at the center of the training paradigm, benefits radiate to all levels.

At the intersection of the three dimensions stands the common element around which specific training revolves, TPM.

The new focus for training has also shifted the responsibility for developing training programs. No longer can this responsibility be placed with each supervisor. It must rest with members of the top management team with full involvement of the TPM coordinating team to assure that it supports the strategic plan.

Much of your training will involve basics or prerequisite training to groom staff for the new TPM environment. TPM training strategy itself must include an introduction to TPM. People need to understand what TPM is in its basic sense: they need to understand it from an orientation awareness perspective before they start learning the fundamental skills and fine points. Next, you'll need to assess whether equipment operators know all they should know about the machinery they are running. Much of the equipment problems we face in our plants today can be traced to the operator—operators unintentionally using the machines improperly because they were never trained adequately in the first place. This problem is not necessarily universal across all manufacturing, but where it exists, it should be corrected through training.

Training is also needed to improve maintenance effectiveness, to ensure that maintenance people are using the technologies and tools effectively. Tools and procedures used effectively are what TPM builds upon.

Another training thrust is to bring the operators and maintenance workers together as a team in small work units in an autonomous mode. For those workers on both the operations end and in maintenance who have been accustomed to "going it alone" for most of their careers, teaching teamwork and autonomous maintenance will take some patience, but it is essential to TPM. Team building, developing people skills, and learning technical skills is a mandatory step toward teaching operators to do routine preventive maintenance. Because operators are in the best position to notice loose bolts, dirty equipment, and low or dirty oil, they are in the best position to tend to fixes for them. Equally, maintenance people are in the best position to teach the operators some of the maintenance skills and practices and learn from them the day-to-day concerns they (the operators) have.

However, underpinning all manufacturing endeavors, including TPM, is the need for leadership to see programs through successfully. Training first-line supervisors to be leaders in a new kind of environment, a natural work unit environment, is another factor critical to TPM success. Despite the TPM philosophy of autonomous teams, employee involvement, and new levels of empowerment, shop-floor activities need shop-floor leadership.

Leadership training is also needed for the small group of people who will lead or champion the TPM effort in your company. This group must be trained to not only champion the TPM effort, but lead the coordination of the program once implemented, ensuring that all involved get the proper training at the proper time. This group would be involved in documenting and measuring the results of the improved equipment effectiveness and preparing progress reports for management.

Your training strategy should also include sessions on actually measuring equipment effectiveness, since improving OEE is what TPM is all about. First you have to devise a way to measure OEE, a way to determine where equipment performance is at a given point in time and measure its step change improvement. Canned courses are available, or you may want to create your own.

Often overlooked in training of any kind is getting everyone on the same communication frequency. That is, talking the same language; it's called process knowledge. People need to understand what the equipment is, what it does, what the major components are and what they are called. A Tower of Babel in TPM won't work. A *frabistat* to a maintenance mechanic may mean something completely different to a machine operator.

· DETERMINING TRAINING NEEDS AND OBJECTIVES

The foundation for determining training needs and objectives is the company strategic plan and the plans that cascade from it. The process for doing it is called a needs analysis. The manner of preparing the needs analysis will differ for each company, but conceptually it might look like this:

1. The desired outcomes of the strategic plan and plans that cascade from it must be understood.
2. The current capabilities of the organization must be assessed in view of the desired outcomes.
3. Organization capabilities needed to obtain desired outcomes must be understood.
4. The difference between current and needed capabilities must be assessed.
5. Desired outcomes that can be influenced by training or education should be identified. Typically, these appear where employee capabilities need to be upgraded. There are outcomes that cannot be helped by training or education and require other actions by management.
6. For each outcome that can be helped by training or education, the following should be understood:
    a. The general nature of the training.
    b. The indicators that show that current knowledge and skills are not satisfactory.
    c. The performance measures related to desired training outcomes.
    d. Other measures that can help in determining if the prospective training is effective.
    e. The breadth and depth of the prospective effort must be understood at all levels of the organization.
    f. The estimated timing and cost for developing and delivering the training. An important part of the cost is the wages of those being trained.
    g. How the training will be used on the job and supported by it must be understood. If there is little use of the training on the job, skills will soon be lost.

7. Survey existing course offerings that meet learning objectives for prospective training.
8. Identify those course offerings that may need to be developed for the prospective training and the learning objectives of each course.
9. Identify specific training needs for groups of people or individuals. Only those employees who will gain needed skills should be trained. In doing this, it is important that the current capabilities of each individual are known. It is also important that any special training needs desired by an employee are considered.
10. Prepare a training plan for intensive review and approval by those involved.
11. After approval, start the execution of the plan.

A TPM TRAINING SESSION

The following description is an example of a TPM training session. Taken from the plan of a company that has implemented TPM, it includes:

- Autonomous maintenance—involvement of the operator in the daily inspection and cleaning of the equipment.
- Product design—considering ease of manufacturing when designing products.
- Equipment effectiveness—establishing a method for gaging equipment performance.
- Productive maintenance—a preventive approach to maintenance involving time-based analysis.
- Equipment design—focusing on enhancing the ease of maintenance and operation when designing equipment.

The training session focuses on autonomous maintenance. After four hours in the classroom learning about the elements of TPM, the group proceeds to the factory floor to clean a designated piece of equipment. The cleaning of the machine is not for housekeeping purposes; rather it is to identify and place tags on items that can create waste. The tagging identifies defective parts, inaccessible areas, sources of contamination, and questions if a part is needed at all.

Participants are given two to four hours to clean and hang tags. On the second day the tagged items are compiled into a list. Team members then prioritize the items and assign responsibilities for completing work on them.

Biweekly follow-up meetings are scheduled until all the tagged items are corrected, and periodic follow-up sessions are held to track the improvements.

CHAPTER 6

TPM in a Union Environment

There are things that in a union environment you can do and things that you cannot do. Any time you have a program that you want to develop, number one, you have to get your employees involved . . . from the shop-floor level, because with involvement there is immediate ownership.

—Kevin M. McGlynn, Facilities Manager,
Briggs & Stratton Corporation

Briggs & Stratton Corporation is a major producer of air-cooled gasoline engines and automotive locks and keys. Their entry into the world of TPM was expedited by the inclusion up front of the International Paper Workers Union, which represents some 6000 company workers in plants in and around the Milwaukee, Wisconsin area. This chapter is an account of the collaboration between company and union that led to their successful TPM implementation, as told by Kevin M. McGlynn, Facilities Manager.

SETTING MUTUAL GOALS/OBJECTIVES, MISSION, VISION

Develop Departmental and Individual Goals

One of the most important factors as you begin to organize TPM teams is to create and communicate goals, both on the department level and the individual level. These goals can be desires from upper management or they can come from middle management or the shop floor. Goals must be mutually agreed upon and communicated to all who will be affected by them. They must of course be approved and, importantly, supported by upper management and in harmony with the organizational goals. They should reflect the divisional goals and objectives. It is these goals that, when properly communicated, will allow your team to properly structure the mission statement and the objectives list. Goals must be attainable, but not too easy as to not challenge your team. Goals must also be measurable. Only through this measuring can it be determined if you have achieved your goals or if you should modify your goals to reflect the results. Set a maximum of four or five goals for your team. Too many goals will dilute the process and too few could provide few challenges to your team. Some examples of team goals might be to analyze all of the equipment within a particular department or achieve a certain percentage of preventive maintenance by a certain date. Remember, they must be measurable and they must be realistic so they can be attained.

Develop a Mission Statement

Your mission statement is important. It reflects your ideals and commitment to your goals. It is from the mission statement that the next list, the team's objectives, is created. The mission statement for the (Briggs & Stratton) Large Engine Division:

"To manufacture engines that meet or exceed customer requirements for quality, delivery, service and responsiveness; to achieve growth and enhance business opportunities by providing competitive value leadership; and to continually improve return on investment above the cost of capital and increase the value of the corporation's shareholders' investment."

The mission statement for the Maintenance Department of the Large Engine Division is:

"To provide a service that adds value to the division by increasing productivity through team-based total productive maintenance activities that improve equipment reliability, capability, and capacity while promoting a clean and safe work environment. Establish equipment management as a shared responsibility among all functions, departments, and employees in order to protect the corporation's and the division's investment in facilities and equipment."

With the mission statement, anyone reading the statement knows the feelings and ideals of the members of the team. This is important for the success of the team.

List Objectives

Objectives explain how you expect to achieve your goals and how you plan to accomplish your mission. The objectives for our division are:

"To serve the lawn and garden and industrial markets with 5- to 18-horsepower, high-volume, standardized engines. In the markets served, it is the intent to achieve and maintain a superior position in customer relations, service, delivery, engine performance, and quality and an equivalent position to the competition in flexibility, lead time, price, and features."

Objectives allow your team the opportunity to express their strategies and tactics. Strategies are the conversion of your goals into action, and tactics are the methods you will use to achieve the strategies. These, too, are important to the success of the team. They also provide the management tools to measure progress.

CREATE A PLAN

Solicit Union Involvement Early

For any program to succeed in a union environment, it must have union involvement and input from the beginning. People do not resist change (as is popularly thought), they resist being changed. If it is the corporate culture in your company to solicit union involvement, this step will be easy. If it is not the norm, you will be pleasantly surprised with the assistance you receive. People desire to excel and improve. They want to contribute and have their contributions acknowledged and appreciated. Explain your vision to the union management and solicit their input.

(We decided to create a preventive maintenance development team [PM-Team]. Its primary mission is to analyze 2400 components in our Large Engine Division and create a PM procedure for each component. The way in which we selected our team members was critical to the success of the program.)

Ask for volunteers to become members of your team. Identify your plans to all of the union members, and be honest. Nothing will make you lose credibility faster than withholding information or being dishonest with the union employees. Remember, for the vision to become a reality and the objectives to be achieved, you *must* have union acceptance and involvement. To receive the acceptance, you must communicate your vision properly and look for employees' input. Do not sidestep this process.

Involve Top Union and Company Management

You can do all of the communicating you want with your hourly employees but without top union management and top company management support, your program won't get off the ground. If you involve the two groups from the beginning, you will receive the support and advice you need to begin the process. Top management officials of both groups will provide information and identify possible problems that might arise from the plan. They can also suggest any contractual changes that might be required for your plan to succeed. Special letters of intent or agreements might be required to proceed with your vision. Early involvement can assist in securing these items. Union/management involvement can also provide some new objectives for you to add to your list. They probably share your vision either in part or in total. You may be surprised and pleased with the support but, if the support isn't there, you can modify your vision or the path you take to realize your vision.

Take Union Officials to Successful Sites

We visited a couple of successful manufacturing sites with our prospective team and with the union representatives. We also entertained companies desiring to see our successful team in action. It is important to pick at least one union-represented company for your group to visit. They should have an opportunity to talk with the members of the maintenance team, ask them questions about their team and the part they play. It's best not to attempt visits with competing companies. Explain the objectives of your visit both to the team you take and to the team you are visiting. Tell your team to take copious notes and ask a lot of questions. Point out, during the visit, the highlights you see and the important observations you make. End the visit with a question-and-answer period followed by an invitation to your facility, once your program gets off the ground. Shared technology between noncompeting industries is critical to the success of any program. Following the successful site visit, set up a meeting where you can debrief your team and brainstorm your findings. Use this opportunity to instill a common vision and a desire to begin the process immediately.

Work on the Plan Together

The "plan" in this context is simply your vision and your goals and objectives. Following the successful site visits, you can solicit input from the attendees to assist you in the development of the plan. This is the first step in union involvement and buy-in. To solicit union involvement in the creation of the ultimate plan, a brief training session will be required. All involved must thoroughly understand the process used in the creation of a mission statement along with goals and objectives. This is the first opportunity to ensure union buy-in. It is important and shouldn't be overlooked.

Implement the Plan Together

It is important that you and the union representatives communicate the plan together. If this is done, walls and barriers will be broken down in the eyes of what is referred to as the "rank and file" (the union membership). The way we accomplished this was to hold informal meetings on company time for any interested parties. The maintenance union steward, the union representative for the maintenance department, was a partner in communicating our plans to employees. We attempted to answer all questions pertaining to the plan. We also presented the proposed timetables, along with the staffing we were looking for. Several questions were raised and answered in a timely manner. We then solicited participation from the group. We wanted to select only those individuals who were interested in serving on a team.

It is important that you base your plan with your corporate culture in mind. If you have a positive working relationship with your union, you will experience few problems, as continuous improvement will be second nature to the members. If you have an adversarial relationship, proceed slowly and with great care. You can and will impact on the union representatives, especially when you visit the successful facilities.

It is important that you stress that seniority will be honored but after the team is selected and trained, it is imperative to keep those individuals on the team. Any time a member is "bumped off" the team by a higher seniority employee, the continuity of the team will be upset for a long period of time. A rapport is developed among the team members and, based on that rapport, anything that upsets the team will affect the progress and could impede the team's success.

Only you and your employees know the corporate culture at your facility. One fear the union workers will have is that management is trying another "program of the month." When you begin this process, you must see it through. You must live it and breathe it. You must be totally committed to the plan and to its success. Union employees will see this and will believe in you and, as a result, they will want to become a part of the team. Bobby Knight, the basketball coach at Indiana University, once equated commitment to a bacon and eggs breakfast: Although the chicken did contribute the eggs, it was the pig who exhibited the greatest commitment.

Throughout the planning process, you must keep an open mind and remain flexible. You will be tested. You can develop all of the rapport you need to get the process off the ground but, if you lose your composure just once, you risk all the credibility you attempted to build. Minds are like parachutes. They only function when they are open. Always keep a positive attitude and an open mind when attempting a process like this.

GET THE PROGRAM STARTED

Select Interested Individuals

The next step is to select the individuals who will comprise your team. The plan is now established; you've held the employee meetings and communicated your plan along with union representation; you have a list of potential interested individuals and you are ready to begin the selection process. One key ingredient is to remain up front and honest with interested individuals. Attempt to explain your plan and answer any questions. You probably will ask for a commitment from the participants. We asked for a one-year commitment from the selected members. We did this to prevent the reeducating of new team members during the one-year period. Obviously, if we had an individual who wasn't interested in remaining and it was within the first year, we would let this individual off the team. TPM requires individuals who want to be on the team. You must remain sincere with your interested individuals. They are making a commitment based on trust in you and your vision.

Make sure you communicate your goals and objectives, your mission and your vision. This cannot be stressed enough; make sure your team knows you are committed to this plan and will fully support it. They must know what is expected of them both from a training perspective and in performing the actual tasks and following the timeline for the plan. Communication is the key.

Again, take the interested parties to some successful sites where they can see firsthand what to expect. They should have an opportunity to discuss TPM with the shop-floor employees there. Invite your union representative; such involvement will continue to build credibility with your team. It will also display the union support you need to ensure success. Take your interested individuals out for lunch after the tour so you can debrief and brainstorm with the team about what they saw.

Announce the Selected Individuals

You have now taken the interested individuals to a successful facility where they had an opportunity to see firsthand what can be done in an industry similar to yours.

You have also communicated with these people what is expected of them. You have assured them you would support the team. You can now announce the names of the selected individuals. You can also send personal letters thanking the interested individuals who were not selected, along with a generic thank-you letter to all of the individuals who showed an interest in the beginning but decided not to pursue involvement with the team. The announcement should be positive and encouraging. These individuals will be building the platform for the future of TPM and the company. It is important to instill this sense of pride in your new team members. They are pioneers, as the maintenance paradigm shifts to TPM.

BEGIN THE EDUCATIONAL PROCESS

Create Your Curriculum

You should discuss, either within your organization or with an outside resource firm, your plans for educating your newly formed team. (We chose an outside source to educate our team.) Look for experts who work exclusively with groups such as this and educate them as to how PM programs are created. They should bring the latest technology with them and remain objective as they educate your team. You should develop a curriculum with the resource firm you choose. Assemble your team and discuss the future training curriculum and introduce the resource firm you chose to lead your team in the educational process. You should develop a partnership with this firm where open communication and dialog can be shared. A high level of trust must be developed between you and the firm, because all proprietary information must be kept confidential. The goal in using an outside resource isn't to provide your competition with proprietary information; it is to develop your team into an efficient resource for your company.

Identify Your Needs

Your needs are specific to you and your company. The corporate culture is known only to you and not to the resource you have chosen. You must identify what you expect to achieve from the partnership. The goals and objectives, mission, and vision statements you have developed will provide a clear outline as to where you want to be. You must create a plan indicating the level of involvement you expect the outside resource to take. Such items as duration of the program, the frequency of the resource party's involvement, along with the cost for services provided, must be communicated up front. Communication to the team will also eliminate any surprises and head off future problems: the team must be comfortable with the resource party. Always keep the team in the loop.

Select the Proper Firm to Educate Your Team

The outside resource might be your local technical college (if they provide services of this type) or someone internally that has knowledge in this field and can be neutral in their opinions and decisions. Or it might be a third-party consulting firm that specializes in this area. Our reasons for choosing an outside consultant firm were many. We wanted a neutral party, first and foremost; we wanted to go with a proven champion in the field, one that had a good track record in dealing with a process like this one; we wanted a professional, one that could educate our work force. We insisted on a firm that would bring the latest in technology to the group, and we insisted on zero errors and zero omissions in our PM creations. We also wanted a firm that would assist us in being as consistent as possible. Our goal was to have the PM creation nameless in that the PMs would all look similar in nature.

We toured several of the recommended firms, especially those where the hourly work force was represented by unions. What we saw impressed our team. The synergy developed between the hourly work force as they created the PMs was tremendous. You will find there are not a lot of experts serving the PM market. You will have to look around, but don't give up, they are out there.

Create a Positive Learning Environment

We built a special learning center for our PM team. This center is outfitted with computers, mainframe and IBM-based; whiteboards to provide the brainstorming resources; several bulletin boards for communications; an audio system; an overhead projector with screen for presentations; and a VCR with a monitor. It is air conditioned and well lit, providing a positive work environment. It is designated the PM-Team's room, fostering ownership. A positive environment in which your team can learn is just another ingredient in the recipe of success.

Begin the Educational Process

By now you have established your goals and objectives and written your mission and vision statements. You have selected your team members and the outside resource to educate them. You have created a positive working and learning environment for your team and communicated all the information to upper company management and upper union management. It is time to begin the educational process, time to put away the old way of thinking. You are moving from the reactive to the proactive. You are implementing the change needed to become a world-class company. Use this educational process as a stepping stone for total employee involvement.

SELECT A TARGET AREA

Pick a Target Area that is Challenging and Beneficial

You must identify your target (pilot) area. This area should contain challenging equipment along with the ability to allow you and your team the time required to fully analyze the equipment. This area must allow the newly formed PM-Team the opportunity to display their new talents in a positive environment, along with the ability to create positive responses from corporate and union leadership. The area targeted must allow the team a technological challenge. The equipment located there must be critical equipment that will benefit from the analysis. The area must be run by technicians who will support and perform the PMs created. Begin slowly and in a predetermined area, keeping in mind the PM created must be a dynamic process. Start small but become aggressive. Pick equipment that you feel will enhance the educational process for you and your team. If the area doesn't work out, stay open to change.

Integrate the Target Area into the Educational Process

It is likely that the area you have selected will hold many surprises. Being too close to a problem often makes us blind to the problem; we tend to overlook problems or opportunities for change. It sometimes takes individuals who are not close to the process to recognize problems and identify the required change.

As you implement the program in your target area, remember to fully support the team in its findings. You will be amazed at the quantity and quality of the suggestions for improvements. Do all you can to solicit suggestions. It is also critical to have the suggestions implemented in a timely manner. This will provide the credibility you need to foster continuous improvement and employee involvement.

Be Critical While You are Educating Your Team

The transformation of operations from one of reaction and complacency to one of proaction and prevention will require individual goals and philosophies to be modified. People don't resist change, they resist being changed. If you involve your employees, they likely will surprise you. People want to do what is right. Create the positive environment for seamless change.

You have to continue to challenge your team to be the very best. They must look at the equipment they are analyzing from an outsider's standpoint. They must be critical in their analysis. Facilitators must ensure they are performing the tasks as required and be critical while educating the team. These team members may have played a part in the design and operation of the equipment they are now asked to analyze. Keep them on track. Develop these new philosophies and instill the new culture in the team members. Measure your progress and modify the process based on the outcome. Again, this is critical to success.

Bring Pride to the Team

Now that your educational process is well on its way, you must also develop the group into a team. Simply calling them a team does not make them a team. You must develop a philosophy with your employees. With the team concept, you have the accumulated knowledge of all of the members. Your support network is strengthened, and you can accomplish more collectively.

One of the most positive ways to promote teamwork is to create a reward system for your members. Communicate to your team that they are a special group of individuals and their efforts will be maximized through the collective efforts of the entire group. Special shirts and shop jackets work. At Briggs & Stratton, we chose a specially designed shirt, unique to our group. The shirt can only be worn by our PM-Team members along with our divisional and manufacturing executive vice president. This identifies our team as a special, unique group. We purchased blue shop coats with the employees' names and "PM-Team" listed above the pockets. These served two purposes. First, they protected clothing as team members analyzed equipment on the shop floor. Secondly, they instilled the sense of unity and identification in members of the PM-Team.

Company publications carried articles about the PM-Team and accomplishments. It is important to communicate to the rest of the facility exactly what the PM-Team is all about. It also provides the positive feedback required to promote the positive work attitude within the team, along with displaying the team's accomplishments.

About two months into our educational process, we had the opportunity to make a presentation to upper management, including the CEO and Chairman of the Board. All team members had an opportunity to voluntarily make a testimonial in this presentation. The feedback we received was tremendous. The support we have been blessed with from our senior management and our upper union management will continue to ensure the team's success.

Announce Your Successes

We have announced our success in several ways. All of the successes are communicated in the form of a monthly report to upper management. This serves as the communication to provide visibility to the affected areas. The reports are not always positive; problem areas should be identified along with successes. Through a positive communication system, all of the individuals that should have access to the information will have access to the information. This can be accomplished through electronic mail systems, internally posted memorandums, verbal communication to groups in meetings, and through visual controls on bulletin or status boards.

Perform the PMs

When our team recommends an improvement, a work order is created and our maintenance department takes the appropriate action. At times we are forced, due to production constraints, to schedule these modifications for a planned downtime such as a summer shutdown.

Up-front planning is critical to ensure that the corrective action or the modifications are performed properly and that the materials required to perform the task are ordered and available when needed. The name of the key contact from the team should appear on the work order. Any sketches or layouts that exist should be attached to the work order.

When the task is ready to be performed, communication to the tradesperson scheduled to perform the job is essential. The reason for performing the modification should be communicated. Exactly what is expected and the time frame to complete this task should also be communicated. Good follow-through practices should be observed.

We are also prepared for PMs as they are created. Most companies use a computerized maintenance management system (or CMMS) for their scheduling, equipment, and materials management. The PMs the team creates should be entered into the CMMS, if available. If not, there are a lot of good stand-alone software packages for plantwide maintenance management of equipment and materials along with labor and cost tracking.

After it is entered into the CMMS, the PM can be scheduled. There are two ways to look at the scheduling process. You can take the equipment out of service and perform the entire PM at one time or you can do as much as possible while the machine is operational and limit your downtime PM procedures to the daily available planned downtime. We chose the second option.

The goal for our PM-Team is to create PMs where a minimum of 80 percent will be accomplished safely while the equipment is in operation during production. The remaining 20 percent will be planned or scheduled for any available downtime.

Our guiding principle is that we must add value to the manufacturing and assembly operations in everything we do. Shutting equipment down to perform a preventive maintenance procedure if that equipment is needed for production will not add value and probably will not take place. In many industries, production control and maintenance seem to be at odds as they attempt to gain custody of the equipment.

Your production requirements and corporate culture must be present in your PM plans. Solicit input from the tradespeople performing the PMs to get the information; they can contribute to the PMs regarding modifications required and their content. They should also provide feedback as to the findings of the equipment. Listen to them carefully and use their input.

LET THE TEAM GO

Provide Support, but Let the Team Analyze

In the beginning, you must provide a high level of support and maintain high visibility. Eventually, the goal is to create a self-directing work team where the objectives are developed and administered by the team. Your goal is to develop the team so they can get to this goal. The leader must participate in the educational process and in the PM analysis as a facilitator or coach. The leader should review all of the PMs created to eliminate errors and any inconsistencies. The team's goal should be to provide zero errors and zero omissions, and none of the PMs should reflect a particular individual's trademark.

Self-directed work teams will lead to the success you want and yield a high level of employee involvement. Lead by example and try to bring the best out of all team members. Concepts such as effective time management and communication skills should be encouraged. You should educate your team in these areas so they can manage the day-to-day operations on their own.

Turn the Analysis into a Learning Experience

Measure results and modify your process based on the results. This, too, is critical to success. It's necessary to make everything you do a learning experience. Down the line, the accumulated knowledge will pay off handsomely and continue to build. Continuous learning applied is continuous improvement.

Celebrate When You Hit a Milestone

Celebrate all milestone events. One good way is to buy lunch for the team. Another might be to plan a trip to another firm. If your contract allows, you might give an afternoon off with pay. Or you might allow your team a day away from the plant to attend a trade show or a conference.

INVOLVE ALL TECHNOLOGIES

Total productive maintenance is one of the most powerful forces in maintenance today. It not only elevates the maintenance profession, it changes corporate culture from one of reaction to one of prevention. There is a problem, however. In today's environment of JIT, cellular manufacturing, pull system versus the push system, and one-part flow, preventive maintenance isn't enough. The equipment must run when we schedule it to run. It can't break down. We are at war with unplanned downtime.

Preventive measures will assist us in the fight against unplanned downtime, but simply using preventive technologies only provides part of the solution. We must bring the other modalities into the picture.

Predictive and proactive technologies are a must in today's industrial environment. These nondestructive and noninterruptive technologies provide a high level of value-added service to the equipment. They provide a baseline to use as the cornerstone of equipment maintenance management. When incorporated with preventive and total productive maintenance, these tools will ensure a successful maintenance process; you will have a high level of confidence that your equipment will run when it is needed without unexpected breakdowns. These tools will also revitalize your maintenance staff. You can eliminate the so called "part-time," full-time repair employee—the person hired to take care of breakdowns who spends most of his or her time looking for something to do. Bring your maintenance staff into the 21st century.

Keep an Open Mind

Remember to keep an open mind. Someone once said, "Winners (successful business managers) must learn to relish change with the same enthusiasm and energy that we have resisted it in the past!" We tend to be comfortable with that which exists. Reactive maintenance is easy work, inasmuch as your day is controlled by outside factors. If a machine breaks, you repair it. Your goal should be to add value in everything you do; if you don't, you transform yourself from an asset to a liability. Keeping an open mind keeps attitudes positive and minds focused on the goals and objectives, mission, and vision.

Implement a Dynamic Process

Constantly review the status and measure successes and failures. This is the only way to ensure that the process is following the goals and objectives, mission, and vision statements developed. Keeping the process dynamic means keeping the process active, able to change to meet the needs of your team.

Take the Process and Implement it in Other Areas

Modify your process as needed, based on the data you have accumulated. At Briggs & Stratton, we placed measurable tools in place. We know we have to have a dynamic process, so if the process adds value, we expand it into additional areas. In your shop, select areas that will benefit the most from the expansion. Use the skilled trades individuals from the areas you are evaluating. Setup people have a wealth of knowledge about their departments and individual pieces of equipment. Seek their input. Listen to them. Learn from them. You and the team can benefit from them.

MEASURE RESULTS

Someone once said, "If you can't measure it, you can't manage it! If you don't measure it, it won't happen." Look at your successes and failures and base your decisions on the results. Successes and failures can be identified only if the process can be measured. Look for opportunities and improvements based on these successes and failures.

Become as Autonomous as Possible

The TPM system must be *your* system. You know your corporate culture and your corporate mission. You know your workers and the level of involvement you can ask of them. These are important variables to consider. Bring a high level of pride to your area through the autonomous interaction inherent in the TPM process.

Ensure Success Through Involvement

The cumulative efforts of many are worth so much more than the efforts of just one. Teamwork can create the positive work environment you need to ensure success. Teams work. It doesn't matter how many resources we throw at a project, without a common focus among all individuals involved, the process will go down the path of decline. Don't develop a process that is destined to fail. Keep all of your employees involved. You will benefit from it. It does take time and effort to develop teams and to foster future development for your teams. By making the union part of the team process, and by providing effective leadership, you give the team its best chance of success.

CHAPTER 7

Predicting the Inevitable and Preventing the Exceptional

Get your maintenance practices in good shape, get your preventive maintenance programs operating, and then expand them to include the operators in the TPM program.

—Wayne A. Vaughn, Facilities Manager, Harley-Davidson Company

PREVENTIVE MAINTENANCE

When implementing the TPM philosophy on the shop floor, good maintenance practices and a thorough preventive maintenance (PM) program are essential. PM helps reduce unplanned downtime and enables the maintenance department to increase machine efficiency through more advanced predictive maintenance techniques. Preventive maintenance can be defined as a system of prevention activities that maintains a machine's proper operating condition through lubrication, inspection, machine overhaul, and part replacement. Maintenance department responsibilities might include coolant recycling, hydraulic oil filtering, and mechanical and electrical PMs. They augment these activities with the predictive maintenance tasks of vibration analysis, oil analysis, coolant analysis, and thermography. Underlying this system is a solid partnership between production and maintenance that supports the program.

On the other hand, operator-based routines, such as cleaning, inspecting, and lubricating PMs form the first line defense against equipment failure, though operators generally perform no more than 20 percent of all preventive maintenance.

Criteria commonly used to determine the level of operator involvement in maintenance activities state that tasks cannot take longer than 60 seconds to complete, tools or ladders cannot be used, and the machine cannot be shut down. Daily startup and shutdown inspections and adjustments—usually documented on checklists—are examples of routines.

Cleaning and inspecting is a practical task in which operators remove wear-causing dirt and grit from machine surfaces, uncover potential problems, and alert maintenance before production is lost. Operator lubrication responsibilities can also include monitoring and maintaining reservoirs at the proper levels, changing lubricants, and greasing.

Job one in getting operators involved is communication. Get instructions out to the operators so that they know clearly what activities are required and when they're required. And make sure the tools they need are available. Operator oiling cards on each of the machines are an effective method. The primary visual control cards on the machines are diagrams of the machines that indicate where lubrication is needed and how often it should be done.

These instruction sheets, or pictorials, are attached to the machine to both identify key focal points and list operator routines and lubrication tasks. Since TPM's ultimate goal is to improve overall equipment effectiveness, zero omissions and zero errors are critical standards in the pictorial's development. To ensure optimum equipment performance, every lube point, every lube type, and every lube task must be identified.

Over time, operators build an intuitive "feel" for their equipment. By learning how their machines function and how to detect abnormal conditions, operators can control the factors affecting equipment performance.

Implementing an effective PM program involves the entire organization and, depending on the size of the facility, will take six months to a year to accomplish.

ORGANIZING EQUIPMENT

Equipment Inventory

Currently, most maintenance people do not look at their equipment from a planned maintenance perspective, committed instead to some other point of view such as management ID numbers. Common practice is to take a conglomeration of machinery and say that it represents ID No. 1.2.3.4. It might really represent 4 or 5, and sometimes even 20 different manufacturers of equipment. Different OEMs will give it one number.

On the other hand, some free-standing (or stand-alone) equipment—like conveyors, presses, and mills—have a single asset number, and the majority of support equipment such as heating, ventilating, and air conditioning (HVAC), utilities, and downstream water treatment equipment is not numbered at all. Thus, a true inventory does not exist.

For TPM to be effective, the facility evaluation must reflect reality. It is best to take an inventory of what is owned and develop it from a planned maintenance point of view. For example, some machines have additional or remote hydraulic systems connected via flexible hoses. These should be inventoried. Maintenance also has responsibility for material handling equipment.

When the inventory list of equipment is received, the next important thing to do is separate the list into three basic categories: critical path equipment, support equipment, and building utilities.

Critical Path Equipment. Critical path equipment is generally one-of-a-kind prototypes with no backups. Product flow stops when this equipment is down. Dissecting a facility in relation to critical path equipment automatically organizes your focus.

Support Equipment. Support equipment serves critical path equipment and is found throughout the plant in numbers far outpacing those of critical equipment. Though needed for production, this equipment would not necessarily interrupt product flow or hinder guaranteed delivery or quality parts.

Building Utilities. Focus here is on any equipment supporting production and building environment, such as HVAC units, air compressors, dryers, boiler rooms, and work treatment plants.

This is the simplest way of categorizing a facility without getting bogged down by listing equipment with little or no value to the critical path process. Generally, from a PM perspective, most equipment will fall into these basic categories.

Critical Areas

Each department contains critical equipment. Selecting a few pieces in each department likely will necessitate others to work within the new PM system from the start. Selecting more than 50 pieces of equipment for the pilot project generally causes problems and slows the first stages of development.

When the PM plan is successfully implemented for this critical equipment, an immediate increase in overall equipment reliability will be realized. It makes sense to start in the critical areas because reactive breakdowns cause considerable delays to JIT and attainment of quality.

Once the project starts, the ultimate goal is to keep equipment running 100 percent during production. Production, in turn, has to understand that the PM plan supports a philosophy in which the maintenance department never goes to the production department to ask them for the machine while it is running. By involving one or two key pieces in each department, or in every cell, all involved strive toward a common goal of keeping critical path pieces moving.

PLANNING FOR PREVENTIVE MAINTENANCE

The Plan

The goal of the phase-one plan is to serve as a format with short-term goals that include organizing a phase-two and then a phase-three plan. The phase-one plan serves as a benchmark to measure the OEE.

The structure of the plan has to be developed from the beginning in a way to support future benchmarking within the corporate structure, where multiple facilities exist, and in relation to company competition.

The plan should be designed to provide an avenue to avoid a complete redo and cultural redevelopment.

Auditing the Process

Crucial to your PM plan is a means to audit its performance. The plan should be able to guarantee you a return on your investment, increase the reliability of the machinery from current performance to a much higher level of productivity, and improve quality and throughput. Three questions arise in developing this measure of performance.

Where are we now? What is the current situation in your facility regarding productivity, downtime, and quality?

What do we like? In each one of these areas, highlight the things you like about where you are now. Note the things that have been working that have been developed over the years; where individuals have made improvements, those improvements perhaps have never been recorded.

What do we want to change? With answers to where you are now and what you like, you can chart what you want to change. Suggest that only one or two changes in each category be implemented and put these on a Gantt chart. Assign the team to make those changes.

The best way to answer these questions is to have a short-term plan—the pilot project—that supports the critical path equipment in each department.

The next step is to establish a mid-range goal that goes beyond the pilot project and includes analysis and organization of all other production equipment actually involved in making the product.

Finally, there is the long-term goal, which encompasses all other support equipment and utilities.

Generally, this approach makes sense to a maintenance person. You have answers to the three questions for the pilot project relative to the program organization and management of the equipment lubrication program. It's important that equipment lubrication be put first. If during an audit it is found that lubricating the machinery is at only a 50-percent level and 50 percent have omissions and errors, then a complex electrical PM system implementation is difficult to achieve. At 100 percent, it's almost a totally wasted effort. The machinery is mechanically unreliable owing to lack of lubrication or the lack of lubrication technologies or systems to support the machinery. Supporting other complex PM goals is hardly achievable if the lubrication tasks are not supported at 100 percent.

Therefore, putting time, labor, money and instrumentation to do more complex PMs without getting lubrication organized first is counterproductive. If you knock out a bearing due to lack of oil or grease it's going to knock out a machine, and that's certainly not a good plan.

Determining the Objectives

Many facilities that want to embark on TPM and PM programs borrow objectives from other facilities, failing to capture their own culture. The key to establishing objectives is to determine them against the backdrop of the particular culture and the personalities of the production departments.

Personalities and Culture. Each facility has a different personality profile for production. Some are laid back, some are intense, some have a high turnover of personnel, others boast work forces with longevity. It is necessary to capture the personality of production and actually record that personality; from those observations you can pick up the culture, identify it, and develop the objectives to capture the culture and enhance it. To capture the culture you have to actually analyze the equipment for PM tasks without using the original equipment manufacturers (OEM) manuals. OEM manuals are good resources to enhance your PM concepts, not vice versa.

Analysis of the Equipment

One of the greatest challenges in developing a TPM and PM program is equipment analysis. How does one go about developing a way to analyze multitudes of OEMs (different types of equipment acquired over long periods of time; prototype equipment; extremely complex, one-of-a-kind design type of equipment) and how does the maintenance department ascertain the thinking patterns of all these OEMs?

Developing the Analytical Process. How do you analyze a piece of equipment? What are some of the basics of the analytical process, especially if you're looking at critical path equipment?

Deciphering what you want to analyze is the first thing you have to do. Go through the "what do we own" list, highlighting (1) the machine, (2) location of components, (3) components, and (4) subcomponents.

Don't Use the OEM Manual . . . Yet. The first guideline that should be adopted to assess the technical level of your maintenance department is to set up the requirements for analysis without the use of the OEM manuals. Most OEM manuals are written by individuals not involved in the production process of that equipment and therefore not aware of the operating environment of their equipment. The best they can do is give some guidelines to protect the wear items or guidelines on what they feel should be done for daily upkeep. Each OEM's manual will differ in the way maintenance techniques are described, and those techniques seldom deal with such things as how the equipment will be used in relation to other equipment, parts replacement, and parts management, all of which are essential to effective preventive and predictive maintenance.

Follow the Power. To organize the thinking across all machinery, follow the power sources into the machinery; they are the key factors. For example, in a CNC machine, you'd start out by following all of the components—such as the motor, couplings, ball screws, timing belts, bearings, etc.—until you end up at the piece part. You would follow the power train through and write the PM in relation to where the energy is moving through the machine and bringing everything together to make the part. In this way, the person diagnosing the machine has a pattern of diagnostics and a logical method for completing the tasks.

Diagnostic Evaluation. The diagnostic evaluation of a piece of equipment has to be performed in such a way that the individual doing the diagnostic work clearly understands how energy is transduced, converted from electrical, to mechanical, to pneumatic, to mechanical, back to hydraulic, back to mechanical. Energy makes things move. It moves through limits and prox switches, through the controlling system and finally gives you a part. Performing maintenance without this level of technical savvy penalizes the potential PM system in that it has omissions and errors to a degree that never gives you your goal of 100-percent equipment reliability.

Identify the Components. When all of the components in a sequence of events have been identified, the next challenge is to write the actual PMs for those components.

Start by dissecting the machine, breaking it down into general areas of assemblies and locations of components and subcomponents.

To illustrate, picture a typical plastics injection mold machine. For the hydraulic system, the first component would be the motor. The second would be the coupling; third, the guard over the coupling; fourth, the pump; fifth, the suction strainer coming into the pump; sixth, the pressure filter coming off the pump; seventh, the bypass filter coming back and dumping the oil back to the reservoir. The manifold assembly has all of the valves on it. You'd break this down, and what you're really doing is identifying all of the components in a power sequence.

What you've done by organizing your thinking pattern is identify each component separately and place it as the first word of your task description. Now you would have all the filters in the entire facility as the first word of each task. You can then diagnose how many filters you need to supply as parts replacement, to store, and what kind. Now you can break down the word *filters* into categories; the same with motors (electric, hydraulic, pneumatic, etc.).

What you're actually doing through the diagnostic concept is building yourself a strong, organized approach to parts management. If you don't do that up front, managing parts when someone finds a problem is very difficult because the word *motor* is 15 words into the task and it is hard to read thousands of entries to find out how many motors you have. The diagnostics have to be a sequence of events in relation to energy sources.

Next (and we're not forgetting the OEM manuals or previous work), after everything is said and done, you go back to the OEM manual, asking yourself the question "have I covered their suggestion?" If so, circle it and note that it's covered in your PM. Then look at previous records, if available (for example, previous handwritten records), and do the same. What you're really doing is capturing all of the cultural aspects of the machinery under one umbrella and bringing it down to a common denominator of how you are going to describe the tasks. Instead of using each OEM as your focal point, the OEM is the backup, an enhancer; it comes after you've developed the dialogue. This approach generally gives you a much better way to communicate across multiranges of different manufacturers, from conveyor systems to high-tech electronics, etc., organizing your thinking first and then having the OEM supply the enhancement. Now you are capturing your culture.

Questions to Ask. The diagnostic process raises some basic issues. The task needs a frequency. Components need descriptions, and energy sources must be identified. The task needs a link name. If you describe the component as a motor, you know it's the drive motor for the hydraulic system, so it needs to have a description for a link name. Also needed is an estimated time for the task. It needs a category that tells whether it will be a PM task or a daily upkeep task. Any additional skills in this task—electrical or mechanical skills—will have to be identified. Is the task trend-analysis related? Does it involve infrared, vibration, or oil sampling? How does it work? Is the task in a confined space? Do I need a lockout-tagout for the task?

The ultimate goal with the first diagnostic work is to prepare to evolve into the future for a paperless system so that the PM concepts and tasks are coded and integrated into a bar code reader or similar technology. In this way, you can carry your computer with you to the shop floor. Everything is in the hand-held computer. All the paperwork is eliminated and the computer will eventually write out the work orders for you as you find the problems on the machinery. That's where your PM program needs to get to.

Proper Description and Methods. Often overlooked when describing components is the future support of parts replacement. Parts descriptions are generally recorded in a haphazard way and consequently tie people's hands in other departments when they're trying to decipher how to support parts management with that particular task. Diagnostic work has to support parts management—not in a reactive maintenance mode, but in a scheduled and planned maintenance mode.

If you've organized the equipment and are confident that the individual can actually diagnose a piece of equipment, the next challenge is to organize the diagnostic work so that the tasks are sequenced and detailed so that they can be easily followed and monitored, with evidence that the task has been completed.

Determining frequency of tasks is both critical and complex. Frequency is one of the things you have to figure out on a task-by-task basis: it can be scheduled on a daily, weekly, monthly, quarterly, semiannual, or annual basis. Or you can have a frequency that spans a greater period of time by having it in a PM cycle that could go out as far as three years.

Some people automatically develop the PM system in relation to frequency, based on how often they should repeat the task, and inadvertently lock themselves into a 12-month period. But in a 12-month period, they probably have only 48 weeks to do the task; with vacations factored in, they likely have less time. If there are thousands of tasks, time becomes the enemy and many tasks go undone. Better to consider a cycle over the normal 12 months and put some of the tasks into an 18-, 24-, 36-month scenario.

The component name link is very important. It is necessary to know what the component is linked to. If it's a motor, is it the hydraulic drive motor? Is it the adjusting screw motor for the platen? Or is it the drive motor for the screw drive unit? Each of the motors on a system must be described in writing.

EVALUATING YOUR MAINTENANCE PROGRAM

A well organized PM program will use all practical concepts available to eliminate reactive maintenance and promote scheduled and planned maintenance. Developing the program and organizing the information on that aspect alone tells you what you're able to support in higher technology. Many plants have either a handwritten or verbal work order system. Often record keeping is at a low level, less than 50 percent of work documented. Typically, people can work in the facility and never record where they worked, what they worked on, or how long it took them. Evaluating and documenting your current maintenance program pays real dividends.

Definitions

What constitutes reactive maintenance? Planned? Preventive? Scheduled? As many definitions exist as there are companies trying to come up with definitions. The challenge is to write down *your* definition of what *your* company perceives them to be.

What differentiates scheduled maintenance from planned maintenance? "Scheduled" and "planned" are similar terms, sometimes synonymous. In one facility scheduled maintenance might mean that they are going to schedule maintenance tasks only during two plant shutdowns. Or they're going to schedule tasks that require rebuild of a machine. Or ones that require getting different skill levels together to rebuild a machine.

Planned maintenance could mean that repairs could be planned when the machine normally goes down. In this way, planned maintenance would never be designed around shutting a machine down to repair it or do PM. It has to be designed to support production at 100 percent machine uptime, with repairs planned only during windows of nonproduction.

In some cases where the machinery is used 24 hours a day, 6 days a week, it would be necessary to go to the production department and barter time.

DEVELOPING THE TASK

Determine Type of Maintenance

Is the task PM, daily upkeep, or scheduled maintenance? What is the task, how should the task be programmed into the system?

Each task has to be separated between a running and down scenario. A sound guideline is 75 percent running tasks and 25 percent down tasks.

Task Ownership

If the task doesn't require specific skills there's no need to develop a task that has ownership. If it's a PM task, anyone in the plant can be trained for it. Where possible and practical, be more generic with the ownership concepts so that the PM team can be cross trained to do all generic tasks.

If the task is associated with sampling oil, ultrasound, infrared vibration, or surge comparison testing, make a note of that up front. Even though you don't own the equipment and you might not have the systems in place to support it, you don't want to have to go back and reanalyze the machine.

Task Complexity. The next block of information you need is level of the task.

Parameters to determine levels are:

Level 1. The task can be performed without tools, often while the equipment is running; even if it's down, it can be done without tools.

Level 2. Specific tools or support equipment are needed to help with the task.

Level 3. Data gathered from the machine's gages, instruments, etc., that tells of the machine's status.

Level 4. Predictive tasks. Equipment and testing devices (e.g., vibration probes, infrared inspection cameras) will provide additional information on which to base complexity level. With this information, the maintenance technician can separate the entire PM system by the level of task.

Another thing you definitely want to know is how long each task will take, the actual "wrench time." The objective is to arrive at basic, workable frequency modes of time: half a minute, 2-, 5-, 10-, 15-, and 20-minute tasks. These task times let us know later that if we create a walking route, how long the route should take and if there are too many tasks on the route. Another consideration: can all PM tasks be completed at 100 percent?

Considerations

Safety. You will need to know if the task can be completed safely. Do you need additional support people? Safety issues should be addressed first and for each task.

Is it necessary to lockout and tagout the machine? It's simple for the person doing the diagnostic work to make those decisions because that way you can check the normal lockout/tagout procedures and see if you can enhance them. It is helpful to print out all the lockout/tagout procedures so you know for sure when the window of opportunity during a nonproduction time allows you to get into a machine.

Accountability. The last portion of the task development is to develop a way of accounting for each task. Answers to five basic one-word questions satisfy this requirement.
- Is it *OK?*
- Is it *damaged?*
- Is it *dirty?*
- Is it *leaking?*
- Is it *hot, cold?*

Answers to these questions generate the corrective work orders. By uploading all data into a computer program, you can easily manage each of the tasks. And by making changes globally, your change from a reactive maintenance philosophy to a productive maintenance practice will be smoother. You want to encourage change; it is fundamental to diagnostic work.

Maintenance Records Control

Plan to identify the level of maintenance records control you now have. The goal is to develop the maintenance records to support the parts replacement and management concept, and the management of time relative to PM, scheduled maintenance, and so forth.

Levels. Record keeping is generally at three levels:
1. None: Absolutely no records.
2. Some written documents, possibly in a file cabinet or even a cardboard box.
3. An in-house computerized support system, or an internally generated database, most often a computerized maintenance management system (CMMS).

Identify where you are and determine whether it is practical to purchase or create a CMMS. One of the pitfalls companies fall into is buying a CMMS prior to identifying a plan for maintenance and uncovering cultural needs. They then try unsuccessfully to cram their culture into this computer program. As a result, they cannot generate reports the way they need them. Their culture will not change, nor will the CMMS program. In most cases starting over is the only option. Normally, the people who selected the CMMS cannot accept that change is needed and get themselves locked into supporting reactive maintenance with their CMM system.

Tracking the System

Tracking the system in today's manufacturing environment mandates the use of computers. Only an automated system can process the volume of data necessary to organize the tasks into a route and to find out how long things are going to take.

Development. When you have completed the analysis on a machine and identified all the critical path machines in the plant and the different departments, you're actually setting up a walking route for the maintenance person. The walking route supports operator involvement, and the person walking the route is the liaison to the operator who's servicing the equipment on a daily basis.

To elaborate, assume that the task development on an injection mold machine identifies 50 tasks. Of these, 10 are daily upkeep in nature, and handled by the machine operator. Although termed "daily" upkeep, some of these tasks are normally done every shift, some once a week. And there may be some not requiring tools that would be considered daily upkeep.

Guidelines. To streamline the process, it works well to organize all "daily upkeep" tasks, and daily, weekly, or monthly inspections, and actually perform them once a day. This eliminates the additional confusion for the operator to figure out what week or month it is, let alone which week of the month and should he or she do only a portion of them every day or do all of them every day. It's best to take all of the tasks considered daily upkeep and organize them so that when the operator walks around the machine in a clockwise pattern, he or she determines before the start of the machine what is to be inspected. And because they can be done without tools, in a safe manner, within 60 seconds, and without any climbing, these tasks can be classified "daily." These daily upkeep tasks are taken away from the PM task and accomplished by the operator. Each task must be associated with equipment reliability and part quality.

Types of Routes. Now you're left with 40 tasks, broken down between a running and a down route. Some can be done while the machine is running and some need the machine to be shut down. For example, you might want to take something apart to look at it and while you're there, you will be opening a control panel and looking at the battery backups and the cooling fans inside the control panel. When you actually have to open up the control panel, you might want an electrician to open it.

The crucial decision is, should these tasks be done every day, once a week, or when? Organizing patterns of tasks or skills is very important. For the most part, on critical path machinery it's best to develop a walking route. That means that the first machine on the route is No. 1, the second is No. 2, and so on. Thus, the PM task orders are printed out in a walking route system. If you happen to be called away on your tenth machine, you can look at your schedule and go back to the eleventh and continue the pattern of walking through the plant.

As you expand the PM into support equipment and finally organize all the equipment in the plant, you'll have what might be called a "mega" route. That route would cause you and everyone supporting it to walk the plant in an organized fashion so that it makes sense.

If you multiply the time by the frequency on that route, you are able to develop a route that is reasonable to walk in an hour. But you have to also factor in lunch and breaks, so what is your actual route time? You might have a route that only represents five hours of actual PM tasks. Your computer program should be able to schedule and track how that's taking place and print in advance the development of the route, what it looks like, and if it is feasible.

Considerations. Before you implement the PM program, you should know the frequencies, the workload, all the lubrication points, how much product by volume you need. These will all be tracked prior to the implementation of the actual PM.

After determining all the various types of data you need to track, you can make a more intelligent decision as to how much of a computer (CMMS) investment you need, what types of reports you need, separate or networked computers, and types of printers. The nature of the PM system drives the configuration of the computer system, not vice versa.

Look to the Future

Predictive maintenance measures physical parameters against a known engineering limit to detect, analyze, and correct equipment problems before production losses occur. It can reveal the cause of a problem, and then, through planned adjustments, help prevent equipment failure and deterioration.

Typically the maintenance department has at its disposal several different types of analytical instruments, for example, an infrared gun, a thermographic camera, or a vibration monitor. Someone generally claims ownership of them and because they claim ownership, other maintenance people don't have access to them.

What is necessary is to list specifically all analytical instruments and support equipment that you own right up front. Answer the questions, how many filter buggies do we own? How many automatic grease guns? How many lube support stations? Armed with the existing instrument and equipment list, you can develop a predictability pilot project and identify any new additional equipment needed to support it.

Even if you don't have infrared or vibration equipment, learn to diagnose machinery and automatically record the need for such equipment for those components that would likely suffer without it. Inventory and justification for analytical support equipment is vitally important to predictive maintenance.

One of the shortfalls in current analysis practice is that the team doing the analytical work does not project itself into the future and actually prescribe a trend test during the first phase. Trend analysis provides the greatest return if it is done up front.

The bottom line is this: in your diagnostic work, record all the information you need for future predictive instrumentation.

Diagnostic Tool Needs

In the selection of diagnostic tools, be aware that technology development advances almost exponentially. This is perhaps more true in electronic technology than in other areas. Continue to bear in mind that what you select today may be obsolete by the time it gets into the plant. However, this does not mean the tools are useless. Generally all it means is that a more advanced model has been introduced.

The most important step in trend analysis is that as you gather information from trend testing, make that information part of a trend concept evaluation; for example, when you find debris in oil from oil analysis, immediately do a vibration test and oil analyses on your pumps to see if there are potential problems. Conduct an infrared test, a visual inspection. Then incorporate that data into a single report to evaluate your plans for scheduled repairs.

LUBRICATION CONSIDERATIONS

Lubrication is one of the most critical elements of a PM program, yet attention to its importance persists at a low level. Common on the shop floor is a lack of knowledge of how to lubricate (the machinery) and a lack of knowledge of all that needs to be lubricated.

Examples

Many machines have greasable couplings, under guards, and some are nonlubricated. Reservoirs on some systems have two worm drives, a primary small worm and a secondary large one: the two are actually bolted together, making it look like one worm drive. The maintenance department never realizes that the first worm has a separate reservoir and a separate grease fitting and the second reservoir has its own reservoir. The OEMs put a level gage on one and not the other, giving the illusion of just one reservoir. This type of understanding or, more correctly, misunderstanding about machinery is the grist of reactive maintenance.

Surface Protection. Most equipment in the facility has basically five surfaces/features to diagnose in terms of lubricant protection:
- A turned surface.
- A ground surface.
- A honed surface.
- A shaft.
- A polished surface.

These are basically the combinations of metal to metal or other materials; their makeup mandates the starting point as to the type of lubricant to use.

The Survey

When developing a lubrication program, first make a floor plan of the plant. Take a top view of the plant and walk through it in relation to either lines, columns, or departments and write down every single lubricant, whether it's in a toothpaste tube, a small can, an aerosol, 5-gallon can, 55-gallon drum, 600-gallon tote, or bulk storage. Note the location, the type of lubricant, the vendor's name, and a slight description of why it is used in that general area.

If you put that information on a spreadsheet program, you'll have a quick summary of what's sitting on the floor in relation to inventory and dollars, and the general lubricants in each zone. Knowing that in advance helps you develop the operator support lubrication stations so that you can evolve putting a lube station in with the correct lubricants in the general area. You need to know this information in advance.

Next, you must organize lubrication concepts and harmonize purchasing with the maintenance department. Too often what prevails on the floor in the facility is that records never agree with those in the purchasing department. In other words, the purchasing department owns records to indicate they have 4 or 5 suppliers and they might supply 20 lubricants. But trying to match these with what is actually on the floor is futile, because some departments have the authority to buy lubricants on their own or lubricants sneak into the plant that aren't needed. That's part of the culture that must be changed.

So, organizing lubrication concepts from what's on the floor for the direct needs of the machinery and matching that and developing one system with purchasing is critical. Studying the records and determining how much you bought from what vendor gives you greater buying clout and organizing the lubricants and developing the PM system in a way not to show the brand or make of lubricant on your records is paramount. In this way, if purchasing finds that they could rightfully change to an equal product and save a dollar a gallon, they're going to change. It's the economic way to go, but it will mean that thousands of records will have to be changed up front. If you use a brand name on your PM records, create a company generic lube code.

Use of Codes

It is good practice to make up a uniform generic code for your plant. Lubricant product charts are available that provide complete breakdowns so you can cross-reference most lubricants.

Removal and Recycling

A good mindset to get into is first, never change oil, and secondly, find a way to never ship the oil out of the plant. That's a sensitive challenge, because it puts pressure on the maintenance department, especially if they're accustomed to reactive maintenance, yet it is key to developing an effective predictive maintenance system.

What occurs too often is virtually dumping the evidence down the drain. Disposing of oil also disposes of the evidence in the oil. The data is critical to machine wear analysis and plotting predictive maintenance. Instead of using the oil to predict premature failure, many maintenance departments go in and change oil, which means disposing of the old oil. That's a waste. By taking a sample of the oil, cleaning it up in the reservoir, and looping it through a filter system, an effective strategy can be drawn as to how to repair the damaged components or change them when the equipment is not running.

Reuse and Additives

Another factor to consider when you *do* have to change oil is the possibility of recycling it through your plant as a different type of product. Most oil suppliers welcome the opportunity to supply additives or suggest a way for you to filter the oil in your plant to bring it up to acceptable cleanliness levels. They can also help in checking the additive level and viscosity. There also is the potential of using hydraulic and gear oils to make cutting oil for machine shop equipment or some regular type of turning or milling operations. This has worked well for many manufacturers. The oil supplier can recommend an additive like sulphur chlorine and you could spike up your own oil, loop it through your system, and have a support, or mix it 50-50 with a cutting oil.

Consider the Downstream Process

In many plants, equipment is largely organized around the manufacturing process, not the downstream process.

Putting together an effective PM system on critical path equipment only will not guarantee delivery of the product to the customer if the downstream support equipment is out of the system. Stabilizing support equipment first is generally not critical path from a maintenance perspective, but stabilizing the environment of the supply upstream and downstream will ensure OEE to the critical path.

Lubricant Selection

The lubrication portion of the PM system must be at 100 percent with zero omissions and zero errors. That means that in the PM system not only do you have to identify all of the lubricants that you actually apply to the machinery—the greases and the oils or any specialty products—but you also have to identify all of the supporting systems that link the loop.

Status Indicators. The level or proximity switches that send signals to the main terminal board identifying a loop fault in the system have to be verified and checked by a trained person to ensure system integrity. If you are looking at lube lines going to toggles and they're turned off, you will verify that the system is mechanically in place, electronically supported, and that all pneumatic actuators on the central loop system are functioning. In this way, you know that the loop systems supplying the product, whether manual or automatic, are all intact 100 percent of the time.

PM Schedule. The PM tasks must be designed in the diagnostic stage to get you to that protection level. This has to be done with zero omissions and zero errors.

Lube Codes and Survey. A generic code is designed to support the immediate development of a PM system. You can start analyzing machinery by picking out the general product in a generic form and eventually incorporating your brand product line. The generic codes can then be developed into your special tags to go on the machinery. The purchasing department can change vendors and product names, but they will always fit this generic code list. Each one of the products that has a generic code, e.g., AW32 hydraulic oil, would have a material safety data sheet (MSDS) supporting it in the general area of maintenance and at each lube station. Also, you would have a primary supplier and a secondary supplier for each product. If the primary for some reason cannot supply you with the product, you have immediate access to a secondary approved supplier.

RESERVOIR MANAGEMENT AT HARLEY-DAVIDSON

With the help of outside consulting engineers, Harley-Davidson developed a simple, yet effective cleaning, inspecting, oiling, and lubrication support system in which operators maintain machine lubrication levels and tradesmen maintain lubrication stations.

Other maintenance department (tradesperson) responsibilities include coolant recycling, hydraulic oil filtering, and mechanical and electrical PMs. They supplement this effort with the predictive maintenance tools of vibration analysis, oil analysis, coolant analysis, and thermography. The foundation of this effort is a solid partnership between production and maintenance that supports Harley-Davidson's drive toward TPM.

Part of the criteria is to make sure that the operator has the resources of lubricants and the equipment, such as grease guns and oil cans, available to him or her. If that's not there it's not going to get done. If he or she has to walk farther than 30 feet, generally it is not efficient from a maintenance perspective, or from a whole system perspective.

Lube stations can be placed by calculating a 60-foot diameter from a specific point and identifying all the equipment within that circle. This data can then be used to determine all the oil and grease types needed to support the area.

The oil containers are color-coded yellow . . . tan . . . white . . . orange . . . black . . . and brown—each a generic, plantwide code that corresponds to a particular H-D number.

PMs are process-sheet printouts that identify daily, weekly, or quarterly preventive maintenance duties to be performed by the maintenance department. These can range from simple, non-interruptive lubrication tasks to part replacements and major overhauls that require scheduling equipment downtime. Pictorials clearly define the route sequence and identify daily operator duties through a simple color-coding system.

Task orders are written to prevent jumping all over the machinery: the operator focuses on a specific area of the machine. The pictorial has a colored balloon, a little colored circle. If it's red, it's an operator's responsibility.

In addition, small plastic tags are placed at each of the reservoir locations, making it easy for the operator to see where to oil.

Pictorials furnish accessible, practical information and are useful tools that help sustain operator-based lubrication.

General mechanical PMs at Harley-Davidson are extensive, and could involve things like checking backlashes, clearances, torques, belts, or various other things on the machines.

The electrical PMs are similar in nature in that they verify electrical voltage levels and other elements to ensure that the switches work and that the lights all work on the machines.

Through implementation of PMs at Harley-Davidson, downtime has been reduced significantly. The company rarely has a lubrication-related problem where formerly they had five or six a week.

The programs thus far implemented at the company have freed tradespeople to do more of the predictive maintenance tasks and provided the time to analyze concerns and devise solutions.

The bottom line at Harley-Davidson is that by integrating preventive and predictive maintenance techniques and folding them into a comprehensive TPM program, overall repair costs have plummeted while profitability and OEE have risen to world-class standards.

CHAPTER 8

CMMS—The Common Denominator

As your company moves toward TPM . . . the organization's ability to respond to and service and repair equipment will rise dramatically. So, too, will the amount of data needed to perform failure and equipment analysis. This highlights the need for a CMMS to track and trend equipment histories, planned work, the preventive maintenance program, maintenance and inventory of spare parts, and the training and skill levels of the autonomous teams.

—Terry Wireman, Integrated Systems, Inc.

Once the new company culture is established, training has been completed, and good preventive and predictive maintenance practices are in place, the addition of a computerized maintenance management system, or CMMS, can enhance proactive maintenance and support continuous improvement efforts.

This chapter, written (with the exception of the AT&T section) by nationally recognized TPM educator and consultant Terry Wireman, describes the CMMS, a database that coordinates maintenance activities and helps identify areas for improvement. Systems range from those residing in simple personal computers to others driven by high-powered minicomputers and mainframes. Although more powerful systems can manipulate and analyze data to a greater degree, any system that organizes equipment, controls inventory, maintains PM schedules, issues work orders, and generates reports will be effective.

MAINTENANCE IN THE MODULAR MODE

The CMMS's most basic function is to organize all equipment information into a workable database. Software has evolved over time from a basic maintenance notepad to a tool that can be used to manage the equipment information necessary to support TPM. Figure 8-1 highlights the main

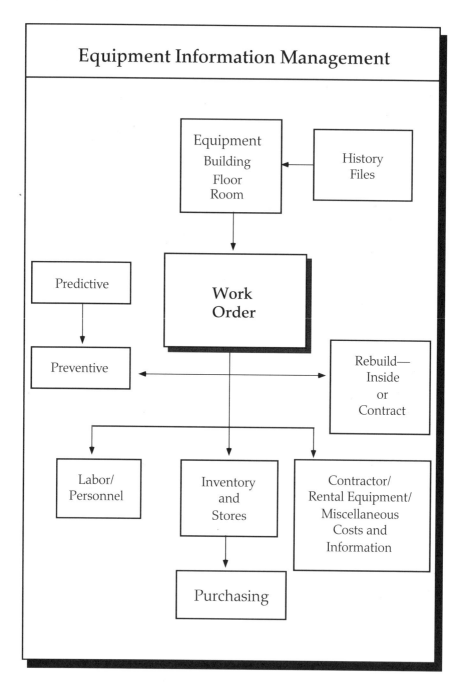

Figure 8-1. Collectively, the CMMS modules organize TPM into an automated system accessible to all workers directly involved in equipment and facilities maintenance.

modules of a CMMS. To understand the full scope of CMMS is to understand the basic function of each of the modules.

The Modules

The PM module is the point of embarkation for most companies implementing TPM. CMMS's main PM functions are to schedule PMs, provide access to PM procedures, and distribute current information to those performing the tasks. The PM module will allow a PM inspection or service to be scheduled for any piece of equipment. In most cases, the PM is scheduled by some calendar setting, such as daily, weekly, or monthly. More advanced systems will allow PMs to be scheduled by some form of usage parameter, such as hours operated, gallons pumped, parts processed, cycles performed. The most advanced CMMS will have the ability to do either calendar *or* usage-based PM scheduling.

The predictive maintenance module (PDM) feeds inputs into the PM module, triggering a PM inspection or service to correct a problem detected by the PDM technology being used, the most common of which is vibration analysis. In a typical scenario, when the vibration analyzer detects a problem with a piece of equipment, the alarm reading from the analyzer sends a signal to the PM module. Based on the parameter set, the PM module triggers a work order that will be used to send a technician to the equipment for further inspection or repair.

The labor/personnel module tracks all of the personnel information on workers who will be performing maintenance. They may be maintenance, operations, or team personnel. Labor rates, personnel schedules, skill levels, and training records are all part of the labor/personnel module. This data stored in a single location on line ensures that the personnel with the proper training and skills will be scheduled on the required equipment maintenance task.

The inventory and stores module enables the tracking of the proper spare parts for the equipment. This includes the physical location of the parts, quantity on hand, economic order quantity, and part usage patterns. With this tracking, availability of spare parts when they are needed is assured. This part of the CMMS is also used to track all spare parts costs as they are used on equipment.

Another of the critical CMMS modules is that used to feed the cost information to the inventory and stores module. This purchasing module processes cost and vendor information from the time a material request for a part is made to the time the purchase order is cut and until the material is received and placed in the stores. This module must provide the accuracy necessary to allow stores to optimize its inventory.

The contractor/rental equipment/miscellaneous cost and information module tracks all outside costs that may be used in maintaining the equipment. This completes the labor and material cost information, allowing the CMMS to properly track those costs to each piece of equipment. In some companies, the use of outside contractors is increasing, so this module is important if a company intends to accurately track its equipment maintenance information.

The rebuild module tracks all equipment rebuild costs, whether they are incurred by rebuilds using company labor or incurred as charges from an outside contract shop. The repair history, such as number of times rebuilt, type of problem requiring the rebuild, lead time to be rebuilt, and movement history is typically tracked by this module. This information is essential when tracking true spares and parts cost histories to the equipment.

The work order module is the key to the entire CMMS. It brings the information from the other modules and formats it so the information can be tracked to the appropriate equipment item. For this to happen, an equipment identifier must be assigned and included on the work order. This ensures that the maintenance information will be tracked to the appropriate piece of equipment. The work order module is closely integrated to the rest of the CMMS, so there is no re-keying of information. Each module feeds the required data to the work order to ensure that accurate information is collected for deposit in the history file of the equipment.

The equipment module contains the information related to each specific equipment item in the plant—when it was purchased, its purchase price, current location, and other item-specific information. The equipment history file contains all the work orders that have ever been completed on a given piece of equipment. All the data for the reports on the equipment's year-to-date and life-to-date repair frequencies and cost histories is contained in this module of the CMMS, and decisions necessary for good equipment management are based on that data. The accuracy of this data determines whether the CMMS can be used effectively as an equipment management information system.

CMMS MATURITY

Since the CMMS is actually a tool to facilitate the maturity process, it must logically follow the same line of development. If the CMMS maturity is matched to the organizational maturity (shown in Figure 8-2), we are able to derive the diagram in Figure 8-3. To grasp the role of a CMMS in the

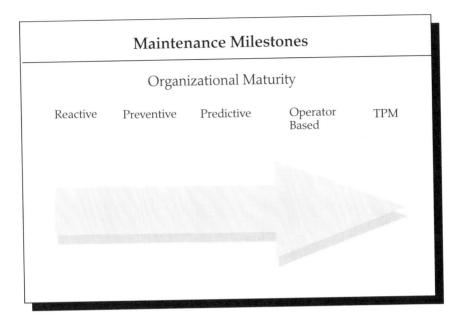

Figure 8-2. Maintenance organizational maturity milestones provide the foundation for CMMS maturity.

TPM environment will require the examination of the maturing of a CMMS usage by following the progression in the figure in the direction of the arrow.

Notepad

When an organization is in a reactive maintenance mode, any computerized system used tracks only important maintenance events, or notes. The integrity, accuracy, and completeness of the data are suspect. Generally speaking, at this stage of maturity only certain personnel in the maintenance organization even care if data is being collected. The maintenance department finds itself so involved in fighting fires, there is little time or inclination to try to input and track complete and accurate equipment maintenance data.

This, of course, is frustrating for knowledgeable maintenance organization people. They know what needs to be done but do not get the resources or the time to properly collect the data. Any data recorded at all will be in the form of a shortened work order, or perhaps some type of notes. In this phase of maturity, few of the features and virtually none of the real functionality of the CMMS are being used.

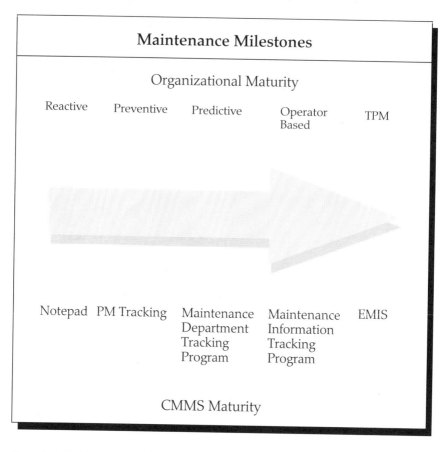

Figure 8-3. CMMS maturity milestones are matched to organizational maturity providing the facilitating tool of TPM.

PM Program

At this stage of organizational maturity, the PM program takes on increased importance. The focus of the CMMS now is preventive maintenance, meaning the PM module will receive extensive usage. As the company progresses through this stage, it will increase the sophistication of its PM program, progressing from a calendar-based program to a usage-based program, and, finally, to a predictive-driven program. The work order module will still be used as a notepad during this stage of maturity. The inventory may be used to identify spare parts, but since most spare-parts information is not sufficiently accurate, it cannot be used to determine cost histories and usage patterns.

Maintenance Department Tracking Program

Most companies today find themselves midway along the maturity arrow. The CMMS is used by the maintenance department, with some input from other departments, to track maintenance costs. While tracking these costs, maintenance specialists begin to plot effective maintenance planning and scheduling strategies. During this phase of maturity, most modules of the CMMS are in use, in varying degrees. Though the data may be incomplete, the company is beginning to understand the value of the maintenance data it needs to pump into the system. The operations department may at this point begin to look at its work backlogs, maintenance budgets, or maintenance work distributions. However, at this stage, the CMMS is still looked at as a "maintenance department" system.

Typically, the maintenance department will show some improvement in its effectiveness, but because the demand for maintenance is still driven by the operations group, no major improvements will be noted. This is due to the operations department still not having a complete understanding of the impact maintenance has on equipment effectiveness.

Maintenance Information Tracking System

This stage of maturity brings other departments into the CMMS fold. The operators may now be involved in directly entering maintenance work requests into the database. In addition, some companies may have the operators performing work, so they may actually be entering and completing work orders in the CMMS and posting the information to the equipment history.

Engineering, too, will be on the system, planning their projects—installation, construction, or overhauls—through the CMMS, thus allowing accurate project history and cost information to be collected. Most organizations within the company begin to base decisions on CMMS data.

In this phase, there is shared ownership of the equipment and also the responsibility for the maintenance of the equipment. A major milestone of this phase is the change from tracking the maintenance department to tracking equipment maintenance information.

Equipment Management Information System

The final stage of CMMS maturity is reached when the CMMS is no longer viewed as a maintenance system, but rather an equipment management information system (EMIS). The information is no longer the sole property of the maintenance organization. Since all organizations in the company use the EMIS, the information now belongs to the company, and since all organizations are involved in entering and using the information its credibility rises exponentially. The indicators used with this information may include mean time to repair (MTTR)—average time to make repairs, mean time between failures (MTBF)—average time between equipment failures, downtime (or uptime), and overall equipment effectiveness.

The indicators derived from the EMIS are reliable and accurate. TPM teams can use this data to focus their efforts to improve equipment effectiveness. When companies have achieved this level of CMMS usage, the ongoing goal of continuous improvement is a reality, not a dream.

TOWARD WORLD CLASS

As your company moves toward TPM and the tip of the maturity arrow, the organization's ability to respond to and service and repair equipment will rise dramatically. So, too, will the amount of data needed to perform failure and engineering analysis. This highlights the need for a CMMS to track and trend equipment histories, planned work, the preventive maintenance program, maintenance and inventory of spare parts, and the training and skill levels of the autonomous teams.

Effective use of the CMMS generates the database necessary to optimize the maintenance resources and help achieve the equipment availability needed to become world class. Each company's database needs will differ, but with well in excess of 300 packages commercially available, manufacturers have many options open to them in selecting a package to suit their specific requirements.

CMMS AT AT&T

At AT&T's 2-million-square-foot Merrimack Valley Works in North Andover, Massachusetts, total productive maintenance is driven by a Unix-based CMMS called MACCS, for Maintenance and Calibration Control System. The company's MACCS is a features, or modules, based system storing information on the plant's 80,000 pieces of equipment. MACCS was designed around specific goals derived from input provided by end users throughout the plant.

Provide Accurate Information Quickly to All who Need it. Plant equipment is expensive; accessibility to maintenance histories and PM schedules helps ensure proper and timely maintenance, which increases uptime and prolongs the useful life of the equipment.

Keep the System Simple. The more comfortable the user is with the system, the more benefit will be derived from it. By making it easier in both the input and output modes, increased use of the system is encouraged. Better data will result because it is easier and more convenient to use.

Reduce the Amount of Software to a Minimum. System dependability and efficiency result from fine-tuning software and streamlining where feasible.

MACCS Modules

MACCS modules were created to meet these goals. This design philosophy led to a continuously improving system from which end users run an average of 1200 reports a month. Users, who range from shop supervisors to product engineers to process engineers, use the reports to help them with decisions about capital improvements, improving machine uptime for better production levels, and time charges and bill backs. Together, the MACCS modules define AT&T's Merrimack Valley TPM program.

Breakdown Records. One module—the most commonly used—is designed to record information on breakdowns. From 1500 to 2000 workers can record breakdowns on equipment in their area. The recorded breakdowns could be anything from a nonworking insertion machine to a tool or a gage, electronic test set, or a problem with the building—a door closer or window stuck open.

Facility Inventories. Another of the modules accommodates the facility inventory, including information about the identified equipment. In this file the user will find information on such items as electronic test sets, where they were built, when they were installed and what they cost at the time of installation, what the current book value is, and the entire known manufacturing, performance, and financial history of the equipment. This module tells where the test sets are located in the plant and which organization "owns" them so that proper charge-back can be made. It also includes information about subcomponents and relationships to other equipment in the area.

Maintenance Schedules. Insertion of periodic maintenance schedules is via another module. These schedules prescribe the frequency of preventive maintenance or periodic calibration, whether weekly, monthly, quarterly, etc. This module guarantees the timely generation of work orders based on schedules resident in the system.

Presently, AT&T uses time-based preventive maintenance schedules. The system has the capability of moving dates, and intervals can be established so that workload can be held to a certain level. Maintenance supervisors, then, can at any time view all PM and calibration work schedules for the year to find out the workload by week and by area for each trade group. This enables them to effectively schedule PMs and calibrations for new equipment as it comes on line.

Work Orders. Where the maintenance tradesperson will spend most of his or her time is in a section called the work order module. The user can log in and view current work pending, from breakdown to a PM to a scheduled calibration. All tasks are categorized and available so the trades specialist can look them up based on who they are assigned to, which organization is responsible for the task, where in the plant the work will take place, and which facility will be involved in the work.

Merrimack Valley's 6500 employees generate approximately 400 work orders a day, coded in three categories: PM for preventive maintenance tasks, PC for calibrations, and BR for machine breakdowns.

Breakdowns are opened by machine operators and dispatched as printed work orders to the appropriate maintenance locations. Planned work orders are generated two weeks in advance of their due dates. The PM schedule references the current document, then routes that information to the appropriate shift, location, and technician. An option of the system allows technicians to print procedures with the work order so that he or she has the entire instruction agenda at his or her disposal. With the procedures are printed the equipment required and any special cautions or lockout tags.

When a technician starts a call, the work order's status progresses from open to start. The MACCS documents changes in the work order status along with the name and electronic signature of the person making those changes. When the work order is closed, the hours charged are calculated and billed to the appropriate shop and the work order is moved from the active file to work order history.

Maintenance Text. Specific maintenance procedures in the MACCS are stored in an area called "Maintenance Text." Maintenance technician and engineer users can call up standard templates and generic documents to guide them step by step through the process of entering data, enabling them to tailor the documents to their specific equipment or usage in the shop. Typically a product or process engineer (now an active member of the TPM team) responsible for the equipment will write a PM document electronically, punch "Enter" and send it to the maintenance organization, where specialists review the procedure and approve it or return it for recommended fine-tuning. It is an iterative, team process involving the "owner" of the equipment and the technicians responsible for maintaining or repairing it. Together they come up with a documented procedure that is workable. Once tested, the procedure is input to the scheduling module where work orders are generated on a periodic basis. The module has the flexibility to accept general information on the equipment and procedures, special instructions on equipment handling, or information based on some history of repairs peculiar to a particular piece of equipment.

All trades specialists have access to this module, both for review and for adding information. With this common text repository and electronic communication, maintenance workers can share information across the plant, across shifts, and across functions.

This module first identifies all equipment numbers associated with that document and contains a general description of the machine. A "Cautions" section alerts technicians to safety-related information, such as lockout/tagout instructions. "Documentation" lists supplemental manuals or drawings on the piece of equipment, and "Theory of Operation" describes how the machine works. If special equipment or tools are needed to perform maintenance or calibration, the information will be listed under "Equipment Required."

A "References" section contains technical support telephone numbers or on-site experts who may be called to help with equipment problems.

Preventive Maintenance Module. The computerized maintenance management system's main preventive maintenance functions are to schedule PMs, provide access to PM procedures, and distribute current information to those performing the tasks.

The AT&T PM module lists the intervals of preventive maintenance, the steps in the PM procedure, and any special information such as special tools or materials needed for the procedure. The system then processes the document automatically by assigning it an ID number that can be used to call up the PM schedule screen.

Information on the PM's status—the shift and maintenance section assigned to perform it, the printer locations to which the schedule is routed, and the technician needed to perform the procedure—is displayed on this screen. If parts kits or consumables are needed, the CMMS issues a parts withdrawal when the PM is scheduled.

Equipment Spare Parts Inventory. Like equipment organization, inventory control requires a large database—in the case of MACCS, a complete parts catalog—to improve maintenance activities.

AT&T's first step was to identify each part in the storeroom, then enter this information into the master parts catalog. This process was initially intended only to expedite locating and distributing parts; however, a serious problem of unclear and inconsistent part descriptions and nomenclature was uncovered: parts were variously described with abbreviations, full names in English, and sometimes simply numbers. The first task of the inventory team was to devise a consistent description paradigm and apply it to all equipment parts in the plant. In so doing, they removed more than 2000 part numbers from the storeroom just by consolidation of duplicates.

Before development of this system, the company's paper-based system was controlling some 25,000 part numbers. Part availability to the shop floor was 89 percent, and nearly 3000 parts were run out of stock. Today, AT&T runs about 400 part numbers out of stock, with 97- to 98-percent availability.

Further enhancing the parts inventory system is an electronic part withdrawal system. Under this system, part requests from any satellite terminal generate order tickets in the storeroom. Clerks can select parts and materials in the order received and package them for pickup within 10 minutes. With each electronic withdrawal, part numbers are recorded in the repaired machine's maintenance text.

The inventory database also enables part identification through equipment number, part number, vendor, or manufacturer searches. These alternative avenues of part identification make it easier and faster for tradespeople to get the information they need to do their jobs.

AT&T has developed further options if these searches fail to yield the desired information. A 6-level noun-tree search utility that progresses from broad to narrow classifications enables the searcher to home in directly on the part. For example, identifying an unknown screw number would start by selecting "common hardware" from a screen of general categories like electronic, bearing, and instruments and gages. This brings up a screen of hardware categories like belts, chains, and gaskets, from which screws is selected. This branches to a list of materials including alloy, brass, nylon, and others. A screen with screw types, such as carriage bolt, lag screw, and slotted flat head appears.

At this point the MACCS has enough information to identify the only bolt of this type in the inventory. The inventory module is also able to track material receipts, requests, issues, and returns—as well as order nonstock parts—and handle reserved parts requests.

Failure Mode Analysis

AT&T's CMMS also supports continuous improvement efforts through failure mode analysis, or FMA. Pinpointing problem areas often begins by sorting work order data. Cross-functional teams—engineers, shop engineers, technicians, and operators—take a close look at facilities that are breaking down the most. Lemons stick out at the top of the list and get the most attention. The teams also find that *words* stick out as well. By analyzing the recurrence of certain words, a pattern emerges. Breaking down the raw data that the operators put in, FMA teams can quickly home in on problems by way of these patterns of words and the frequency of their use.

Once a particular machine has been targeted, a word analysis is conducted through its work order history, which in turn generates a report identifying the most commonly occurring words found in the problem description and solution sections.

In the event a particular machine has a lot of downtime, the FMA team will run a report spanning a 6- or 12-month period to find out what's actually happening to that piece of equipment. The team at this point has options as to the type of analysis it wants to employ. Through the MACCS, they can run all the hard orders in a brief format and go through it manually, or they can conduct a word search. The word search enables them to pick up abnormalities in the form of word repetition. If the word "bearings" shows up frequently, the analyst in real time is able to select the work orders pertaining to bearings, and quickly narrow down the problem area. From that point, it is simply a matter of gathering a team of people—tradespeople, shop supervisor, maintenance engineers, and shop-floor engineers—to analyze the data and devise a solution. The advantage of having FMA on the MACCS is that members of the TPM team can get into the system, rapidly identify the problem area, and come up with a solution without spending a lot of time just trying to locate the trouble. In this way, MACCS does a real yeoman's work in increasing uptime.

To illustrate: the company had a lot of clamp issues with the side clamp on a particular machine. So the word "side" and "clamp" showed up at the top of the distribution. Instead of looking at all components of the machine to identify the problem, the team was immediately directed to the side clamp. From there, they determined why the clamps were failing. They could look at a year's worth of data quickly to find out which items showed up with the greatest frequency.

In another case, when a cable-forming area was analyzed, the team found that the machines with the most reported problems were also causing a bottleneck. Through word analysis it was discovered that the most recurring word was "punch," referring to problems with the punches on the machine.

Further analysis uncovered that these insertion machines used multiple-size connectors. Occasionally they would be set up for connectors of the wrong size, causing the punches to hit against the steel tooling and break. Two solutions resulted. First, the team modified all the programs in the machine. This helped eliminate the connector problem, but there remained the potential for human error in the form of loading the wrong program. This led to the team modifying the physical machine so that if the punches were to fire they wouldn't hit the tooling.

Though operators and tradespeople knew for some time that punches were the problem, it wasn't until the data was collected and studied that the full extent of the problem across three work shifts, different machines, and different shops was recognized. In addition to working with the punch supplier and saving around $30,000 on the cost of punches, modifying the machines to prevent punch breakage reduced the number of punches needed by 75 percent.

CMMS and Cost Accounting

One of the elusive goals in a large corporation with many departments and numerous workers is that of maintaining accuracy in charging time and materials back to internal customers. This function at AT&T prior to MACCS was handled by an electronic time independent system (ETIS) that could never be related back to actual orders. Also, if there were errors in the accounting, they didn't get back to the originating department until a month later. With MACCS, when a tradesperson answers a call and opens and closes a work order, the real time he actually worked on that work order is automatically transferred to the ETIS with the correct charge against the correct department. Maintenance supervisors now only have to log into MACCS once a week and review their people's times to compare time charged to time worked and make any adjustments needed. Those changes flow back to the MACCS system, which talks to the ETIS, ensuring that the customer is getting charged correct time and material rates. If there is any question on the customer's part, the maintenance supervisor can easily log onto the MACCS, call up the orders, find out exactly what was done, what materials it required, and how long it took.

CHAPTER 9

The Equipment Operator as Equipment Owner

> *They become part of the equipment, the equipment becomes part of them. It's theirs. It's not Maintenance's anymore. It is not Magnavox's. It's their equipment and they want to take care of it. Without the operators, you don't have a TPM program.*
>
> —Richard N. Burdek, Manager, Equipment Services,
> Magnavox Electronic Systems Company

One of the most visible departures from the norm that you'll find in a TPM shop is the role equipment operators play in the servicing and minor repair of their machines. Their active and regular involvement in maintenance represents a major shift in responsibility and authority on the shop floor.

This change that TPM brings can be a culture shock to traditional organizations. Employee empowerment, operator ownership, teamwork, and autonomous maintenance will be challenging concepts to supervisors who confuse sharing responsibility with the loss of power and view empowering operators as an encroachment on their domain. This is why it is important in planning for TPM to get supervisors—especially first-line supervisors—involved right from the beginning, to educate them on the advantages of TPM and how TPM will make their jobs easier.

AUTONOMOUS MAINTENANCE

At the core of operator involvement in TPM is something called *autonomous maintenance*. It's the hands-on activity leading to improved overall equipment effectiveness.

Autonomous in this context doesn't mean that maintenance is done in a vacuum. It simply means that certain equipment maintenance is done by a team of operators and maintenance tradespeople who are closely involved in the daily operation of equipment. The team is the focus of such essential daily tasks as cleaning and inspection, lubrication, monitoring, and other tasks formerly under the umbrella of the maintenance department.

It's similar in many respects to owning an automobile. To you, its owner, a car is a major purchase, representing a sizable cash outlay that you'll likely be paying down for some time; therefore it takes on importance in your life. You expect your car to be dependable and provide quality transportation for many years. Hence, you have strong incentive to keep it running in good order. But it's a complex piece of machinery needing special care: cleaning, inspection, periodic service, and occasional repair and new parts. By performing regular service and minor repair and replacement when needed, you can help ensure your car's dependability and longevity: every 3,000 miles or so, you change oil; you fill a tire when it is low; change a bulb if it is burned out; tighten a belt when it is loose—all simple, common-sense-type things that prolong the life of your car.

Equally as significant, by driving your car every day, you as owner/operator are in the best position to detect subtle differences in the vehicle's operation: such things as a new vibration, or a noise that wasn't there yesterday, or a change in the way the car handles. (Ever take your car to a garage and tell the mechanic to listen for that "new" rattle under the hood, only to have it disappear in his or her presence?)

That's precisely the idea behind operator, or autonomous, maintenance. Shop-floor equipment is costly and complex. Just like your car, it needs periodic cleaning, lubrication, inspection, and regular attention. It needs pampering, tweaking, adjusting, and listening to on a daily basis, same as a car. It also needs regularly scheduled (preventive) maintenance just as your car needs a tuneup every 30,000 miles or so.

Unfortunately, the equipment operator usually lacks the feeling of ownership. Conventional practice on the non-TPM plant floor is "I run it—you fix it," or "I'm manufacturing, you're maintenance." And the older the company, the more rooted this mindset becomes and the more difficult it is to change. But to implement TPM, change it must. In a TPM shop, the operator becomes the "owner"—the focus of routine maintenance and the central figure in overall equipment effectiveness.

With TPM properly planned and implemented, operators will more easily take ownership of equipment, and tradespeople will be more likely to let them. The key—once again—is commitment. With the entire organization supporting the TPM philosophy—from top to bottom and from bottom up—the wish to "not get left in the dust" will help turn resistance around. Early, effective, and coordinated training in the values of empowerment, authority, and ownership implicit in team playing should do the rest. The important thing is to get across that management is serious about this thing called TPM and is supporting all its players with the collective clout of the entire company.

ROI: REAL OPERATOR INVOLVEMENT MAKES FOR REAL RETURN ON INVESTMENT

Operators can make or break a TPM program. Without interruption of their production work, they can easily prevent breakdowns, predict failures, and prolong the life of their equipment by becoming more intimately familiar with their machines. But to do this, they have to become highly equipment conscious, and that means intense training. Operators must learn what is normal and abnormal operation—what they should listen for and be alert to. They must know what to do to keep their machines in normal operating condition—lubricate them regularly, monitor vital signs, and record abnormalities. They must also know what to do to get the machine back on line when something goes wrong—if a minor problem occurs, fix it; if major, call in the maintenance member of the TPM team to schedule repair. These "must knows" for equipment operators are not intuitive; they must be learned.

Operators must be taught how to lubricate, when to lubricate, what to lubricate, and the best methods of checking lubrication. If they are not in the habit of cleaning their equipment—which in TPM means inspecting—operators will have to learn how. Simple good housekeeping to keep debris from around machinery is not necessarily a part of the conventional operator's job description, but it is mandatory in TPM. So, too, is changing procedural habits—or developing some—and becoming maintenance conscious.

Over time, through formal training and applying that training on the job, operators will know what to look for in their inspections; they'll begin to look at their equipment from a maintenance point of view as well as from an operator's perspective. They'll learn their machines' critical points and areas of greatest wear potential and the relationships between certain abnormalities and what causes them.

In the course of becoming more familiar with the equipment, a sense of ownership will begin to take hold. With this ownership will come an understanding of the role the equipment plays in producing quality and how this new operator/machine relationship fits with the goals of the company. A new confidence in being able to diagnose problems and devise solutions will emerge. Operators will become more creative in problem solving as they become more comfortable in their new, empowered role.

METAMORPHOSIS AT MAGNAVOX

Just a few short years ago, the predictable response at Magnavox Electronic Systems in Ft. Wayne, Indiana, to an operator doing anything that smacked of maintenance would have been, "You can't do that. You're not qualified. You're not trained for that. You don't know what you're doing." Theirs was a traditional shop; like thousands of other businesses throughout America it functioned within a hierarchical structure, with management providing detailed direction to workers who weren't challenged.

But Magnavox has changed. Through their "TPM Certification" program, teams are achieving autonomous management. They've expanded their preventive maintenance responsibilities, improved their equipment and processes, rearranged their work areas, and been instrumental in the purchase of production equipment.

TPM "champions" at Magnavox began their efforts by briefing the vice presidents and the directors of various operations, describing to them what TPM was and what it could do for the company. To make TPM work, they had to make sure management had full understanding of a philosophy of maintenance that would mean they would have to give up some of their power. Facilitators next went to the supervisory and managerial levels throughout the company. They were the hardest sell: within the TPM structure, they were asked to go from being supervisors to "resource providers." Initial reaction was predictably cool. But when they learned that their jobs likely would be more rewarding because of less stress, supervisors warmed up.

The next step was team building with the operators and maintenance people. Magnavox provided 30 to 50 hours of basic training in the concept of team play, since these workers had never worked as a team before.

Most of Magnavox's manufacturing areas are broken down into manufacturing cells. TPM teams are made up of members of individual cells (operators), a maintenance person, the supervisor, and any manufacturing engineers involved with that cell. If a resident quality person is assigned to the cell, he or she would be a member as well.

An individual team's path to autonomous maintenance, following teamwork training and team formation, starts with gathering equipment data. The teams receive basic TPM training on subjects like baselines and measurements, the OEE, and visual controls. They perform an initial cleaning, identify and resolve defects, and set their own goals and standards. All teams receive the same safety training, but machine setup and operation is unique to each work cell.

Workshops supplement the basic TPM training and help improve equipment and processes by focusing on meeting team goals and standards. Inspections and audits—usually conducted by other teams—determine when a team is certified. While Magnavox endorses this process flow, it is the team that determines their rate of progress. To become certified usually takes from 6 to 18 months, but can take as long as 2 years, depending on the complexity of the work cell and, quite frankly, the level of enthusiasm of the team.

There is no set direction for the teams. Each team sets its goals based on what it thinks it is capable of doing. Recognizing that TPM's success rests on management's ability and willingness to empower its workers, Magnavox continually communicates the team's importance in the improvement process and the operator's role in that process.

Operators are the recognized experts on the machines at Magnavox. They spend the most time with them, and as a team, they make recommendations. It is a cooperative effort between management and the TPM teams. What makes it work is the stable, secure TPM environment that stimulates open communication and nurtures empowered employees.

Operators who once viewed equipment as "just a machine to get the job done" take possession of "their" equipment, the process becomes "their" process, and the time at work even becomes "their" time to decide how to spend most productively. These are the transformations of success.

Magnavox finds that empowered team members usually develop common characteristics. They'll voice concern about their work areas. They'll seek more information on their processes. They'll lose their fear of failure. They'll open up to new methods. And they'll help set team goals. In the process they develop their own methods to reach their standards so they can reach their goals. Goals might include keeping the machine looking like new, or maximizing machine utilization, or zero defects and zero scrap, or performing preventive maintenance and minor repairs.

Empowering team members to schedule and perform PMs is a basic but important autonomous activity. The shift from the directive style of management to teams has changed the entire manufacturing arena: to get something done on the shop floor now, you go to the team, not necessarily the floor supervisor. The teams take action where before all shop-floor business went through the supervisor. If, for instance, engineering has something they want done in a production area, they don't have to go up the ladder and then back down again. The TPM process goes laterally. This not only saves time but brings operators and engineers closer, resulting in a much broader understanding of each others' concerns.

Teams establish cleanliness, lubrication, and operational standards. Once a standard is set, procedures that support it are developed and documented by the team. At Magnavox, team members also review and revise existing maintenance procedures that affect them. During the formation of a team, one of the first things it does is perform an initial cleaning. This is quite important to Magnavox, since some of the operators have been working on a particular machine for 10 to 15 years or more, and in all that time they haven't been permitted to touch any other part of the equipment than the on/off switch.

This initial cleaning serves a couple of good purposes. First, of course, it gets the machine clean so defects can be more easily detected. But, perhaps more importantly, it gets the operator more familiar with the equipment. He or she learns to differentiate the good vibrations from the bad ones. It prompts questions by the operator to the maintenance person. It etches in the mind that cleaning is inspecting; during the initial cleaning the operators do an inspection of the equipment. If there is a screw missing, a part broken, a light burned out, they tag it. In the defect log, team members open work orders with the equipment services (maintenance) department, and track them to ensure that all are resolved.

Fluid and Flexible

Because TPM is a process of continuous process improvement, goals and standards are not fixed. As teams mature, it will be necessary for them to periodically review their goals, standards, and procedures to ensure their purposes are being achieved.

All TPM teams at Magnavox monitor their own equipment and use checklists as daily PM tools. Checklists are useful in collecting, identifying, and analyzing equipment data. But equally important, operators need to use their senses to monitor the equipment's operating condition. Certain checklists identify the areas of machines that need cleaning week to week. Others indicate which gages should be checked. Still others remind operators to check materials, valves, and speed of product throughput.

As operators become increasingly familiar with the checklists, they develop an "operator's sense" of their equipment. If a chain is jumping or pumps are not consistent, he or she will sense it.

With a change in manufacturing processes or machinery comes the necessity to update checklists and PM procedures. It is critical to the TPM process that the team make those changes. In Magnavox's TPM process, if something needs to be updated, the team discusses the suggested change, makes a decision, and one of the team members updates the database on the computer.

Empowerment Engenders Ingenuity

The principle of ownership was amply demonstrated by one of Magnavox's TPM teams upon the occasion of their getting a new piece of equipment. When the machine arrived, members of the operator/maintenance team conducted an initial cleaning and inspection to (1) more fully understand its operation before putting it into production and (2) to get familiar with it from a maintenance point of view. They completely disassembled it, cleaned it, lubricated it, put it back together, and put it on line. The team then had the advantage of knowing the critical parts of the machine and those areas subject to greatest wear. They were able to anticipate the machine's needs in terms of lubrication and parts replacement and contribute knowledgeably to preventive maintenance planning.

Arguably the greatest challenge to certified teams at Magnavox—and to TPM facilitators in general—is to keep momentum going, to continuously improve equipment and production processes. When TPM is first implemented, dramatic results are recorded in process and production figures: 30 to 50 percent jumps in uptime for equipment were recorded and increases of 25 to 30 percent in output were achieved, while rejects were cut as much as 90 percent. Once past these initial dramatic successes—which play well on board room charts—gains come at a slower pace and TPM achievement curves reach a plateau. Maintaining or inching up that plateau now is the new and continuing challenge of the TPM teams. Although management will likely stay happy if the effort sustains the plateau, continuous scrutiny of equipment, processes, procedures, and standards is essential.

Such continuous improvement at Magnavox has produced big results from some seemingly small problems. To address a flux contamination problem in a wave solder machine, the TPM team installed a density controller to increase flux life and added cleaning brushes to the front lip of the flux pot. This saved the company about $16,000.

Further inspection of the same machine resulted in installing "L" fingers to eliminate a problem of rejects caused by parts falling through the conveyor. Other improvements include a sensor in the density controller that triggers a caution light when flux approaches levels that may damage the pump motor.

Team members have also suggested operational requirements and have even become involved in the acquisition of new equipment. Operators were asked for the first time for their input on specifications for new machinery. They toured other companies, discussed with their counterparts attributes and shortcomings of different makers' machines and gained a broader knowledge of available technology. They shared this information with the equipment purchasing team.

On another front, a TPM team organized around a test station restructured their work area to improve process flow. Originally, their cell was in 3 sections and manned by 20 people. Each operator worked out of a 25-board carrier, which might or might not be completed in a work day. The quality person was located in another part of the plant.

The team studied the process, calculated the time spent moving carriers and how long they sat idle before being moved to the test station. From this, they instituted a *kanban* (JIT) system, reduced the team to just four members, and moved the quality person into their area.

Though the team is smaller, the improved process flow has resulted in a higher output of boards. The team's minimum quality standard is 250 boards per day or 1250 per week: the weekly record stands at 1687 units shipped.

The Payback

TPM and autonomous maintenance success at Magnavox can be measured by several yardsticks: the bottom line, to be sure; but also in reduced equipment downtime; shorter setup times; and less quantifiable, but equally important, increased operator enthusiasm, more togetherness, more teamwork.

CHAPTER 10

Putting it all Together

Moving from traditional maintenance and manufacturing practices to TPM can be made much easier by first understanding all of the concepts and strategies associated with TPM in its purest form and then building on your current work culture and equipment management practices.

—Robert M. Williamson, Strategic Work Systems

Throughout this book, we have attempted to communicate a theme that is central to implementing TPM: changing from current ways of maintaining equipment to equipment management and ownership and the careful examination and planning that change requires. TPM expert and educator Robert Williamson summarizes in this chapter the key elements that make TPM work and the changes necessary in corporate culture to nourish them.

TRANSFORMING THE TRADITIONAL MAINTENANCE CULTURE

All too often companies decide to "implement TPM" with a *primary* focus of involving operators in the maintenance of their equipment. In many cases this is the wrong place to start. Others try to follow the Japanese TPM models as a cookbook, with limited success. To be successful—that is, to show significant improvements in equipment effectiveness—TPM must be adapted to suit your company's needs and built on the basics of what is *already* in place.

TPM is an equipment management strategy, not a maintenance management program. The difference may seem subtle, but the results are dramatic. As long as equipment is cared for solely by the maintenance department, it will rarely reach optimum levels of performance, availability, and rates of quality. The highest levels of equipment effectiveness are achievable when the equipment is managed by the people who operate it, maintain it, and design or improve it. This, of course, means a *team*. Figure 10-1 summarizes how to put the "TPM Puzzle" together for transforming the traditional maintenance culture to total productive maintenance.

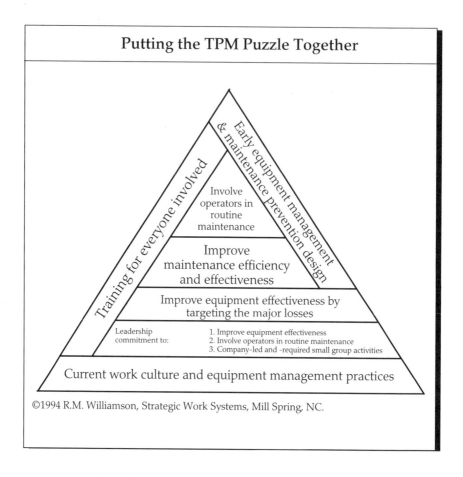

Figure 10-1. TPM begins at the foundation level of what is already in place at your company and builds to the levels of operator involvement and early equipment management and maintenance prevention design. (Courtesy R.M. Williamson, Strategic Work Systems.)

Putting the TPM Puzzle together begins with the foundation: *current work culture and equipment management practices.* All other elements of a successful TPM transformation build on this base.

Current Work Culture and Equipment Management Practices

For people and organizations to embrace TPM, they must be able to see it as improving current practices rather than a radical departure from them. The key point is to *build on the basics of what is already in place.* Go after such critical success factors as:

- Learn as much as you can about TPM. Those who are leading or planning the TPM effort should get as many different perspectives of TPM applications as they can. Many books, conferences, and seminars are available.
- Look for aspects of TPM in your business. Note how successful they are. Look at basic work practices and job descriptions and for ways to link TPM to other world-class initiatives.
- Gather a small group of interested people to explore the potential applications of TPM for your business. (Depending on your labor relations you may or may not wish to involve the union leadership at this point.)
- Identify some equipment-related business results that need improving. Make some initial OEE calculations on the "suspect" equipment/system. Look for bottlenecks, frequent downtime or delays, high maintenance costs, poor or inconsistent rates of quality.
- Conduct a *TPM readiness assessment* to develop a customized approach for TPM. During this assessment, management should announce their intent to explore TPM to improve equipment effectiveness for specific (stated) business needs.
- Build your emerging TPM work culture on elements of TPM that are already in place. Avoid presenting TPM as an entirely new approach.
- Use the basic "Five Pillars (or Elements) of TPM" as a guide.
    1. Improve equipment effectiveness by targeting the major losses.
    2. Improve maintenance efficiency and effectiveness.
    3. Involve operators in the routine maintenance of their equipment.
    4. Train everyone involved.
    5. Design for early equipment management/maintenance prevention.

The "Five Pillars of TPM" must exist to achieve the highest levels of equipment effectiveness. Each of the pillars is a proven method for improving equipment performance. The true TPM advantage comes from the relationships of *all* five being established and continually improved.

Leadership Commitment

Initially, plant management, department management, and labor leaders in unionized plants all must be committed to and supportive of the three main concepts of TPM as they apply to the current work culture and equipment management practices. Ultimately, everyone in the organization wishing to establish a TPM work culture must commit to the same three concepts:

1. Improving equipment effectiveness.
2. Involving operators in routine maintenance.
3. Implementing company-led and -required small group activities.

Improving Equipment Effectiveness

Leadership at all levels of the business must understand the major equipment-related losses and what they mean in terms of *hidden capacity* in the plant. Overall equipment effectiveness is defined by factoring equipment availability, performance efficiency, and rates of quality. In many cases equipment operates at less than 40 percent effectiveness. Improving equipment effectiveness to 80 percent doubles its throughput, or capacity. By continuously improving equipment effectiveness, any investment in new equipment can be postponed or, in some cases, eliminated.

Involving Operators in Routine Maintenance

Historically, operators in many plants ran the equipment, maintenance repaired it, and engineers designed modifications for it and planned new equipment. There was no real sense that anyone "owned" and cared for it, as one would with personal property. Equipment was treated like a rental car: drivers assumed that someone else was maintaining it and, consequently, no one did. They typically ignored the routine inspections they would do on their own cars, and maintenance carped about how those drivers abused the rental cars by running into curbs, pot holes, and generally trashing the vehicle. The parallel between rental cars and shop floor equipment was striking.

But TPM changes all that. Operators *are* involved in the routine care and upkeep of the equipment they operate and monitor. They are part of a larger team that collectively has a sense of *ownership* of the equipment. Leadership must value the role operators play in the routine maintenance of equipment.

Company-led and -required Small Group Activities

Quality circles and total quality have led to problem-solving groups and other ad hoc cross-functional teams to identify or make improvements in the manufacturing process. In most cases these are voluntary groups and their levels of success reflect the vulnerability of volunteerism. TPM, however, requires a completely different type of group.

TPM groups are *equipment-focused natural work groups*. They all share a common bond—the equipment and its desired measurable performance results. Performance will improve if everyone in the natural work group treats the equipment with a sense of ownership. The equipment-focused natural work group consists of all operators from every shift; all maintenance personnel; all equipment and process engineering and technical staff; all supervision and management; and any other groups who directly influence the way equipment operates. Leadership must recognize that the highest levels of equipment effectiveness require the highest levels of equipment-focused teamwork, much like that seen around a competing *NASCAR Winston Cup* race car.

Perpetuating the Three Concepts of TPM

A demonstrated commitment to the three concepts of TPM will communicate that the methods and results of TPM are important to the success of the business. Leaders who can communicate in terms of these concepts and hold others accountable for achieving and sustaining the desired results contribute measurably to a solid TPM work culture. Leadership can continue to reap the benefits of TPM by continually reinforcing these concepts and rewarding the outcomes.

Improve Effectiveness by Targeting Major Losses

TPM work cultures are *data* driven. There's no getting around it. Equipment effectiveness data allows work groups to systematically target and reduce equipment-related losses. Begin your TPM initiatives by looking for opportunities to improve process and equipment performance. Search for *specific* opportunities for improving the major equipment-related losses such as:

Planned Shutdown Losses
- No production scheduled, shift change, breaks.
- Planned maintenance shutdowns.

Unplanned Downtime (Availability) Losses
- Failures or breakdowns.
- Setups and changeover.
- Tooling or part changes.
- Startup and adjustment.

Performance Losses
- Minor stops.
- Reduced speed.

Quality Losses
- Scrap.
- Defects, rework.
- Yield.

To achieve significant bottom-line results, focus the TPM activities on eliminating equipment-related losses. TPM is not an "activity centered" improvement process with results measured by number of teams formed or amount of training conducted. *TPM is an equipment-centered, results-oriented improvement strategy.*

Establish equipment performance metrics, measure performance, and communicate throughout the equipment-focused work group. Pareto charting methods begin to address the root causes of the high-loss items.

Improve Maintenance Efficiency and Effectiveness

At the core of TPM are the basics of maintenance efficiency and effectiveness—preventive maintenance, predictive maintenance, maintenance planning and scheduling, spare parts, lubrication, maintenance management systems, and so forth. All the technologies and methods for improving equipment maintenance are included in this pillar of TPM. RCM, or reliability-centered maintenance, is another method for improving maintenance efficiency and effectiveness in large, complex equipment processes.

Good basic maintenance practices have formed the basis of TPM since its inception. American-style preventive maintenance is credited as the 1950s foundation for productive maintenance which in turn became the foundation for total productive maintenance in Japan in the late 1960s.

When beginning to establish a TPM work culture, make sure maintenance efficiency and effectiveness activities are focused on the major equipment-related losses.

Involve Operators in Routine Maintenance of Equipment

People who operate and monitor equipment in manufacturing, process, and utility industries should pay close attention to the way their equipment performs. They should look for problems that could cause failures or downtime, off-quality production, and slower-than-design operating speeds or efficiencies, much the same way they pay attention to their own cars and trucks. However, do not put tools in the hands of operators until they and their leadership are supportive, maintenance is supportive, and proper training and qualification has taken place.

Involving operators in routine maintenance builds on their sense of ownership of the equipment and the recognition that skilled maintenance and technical people are generally called to address the increasingly more complex equipment problems and tasks. Because operators are often much closer to the equipment than engineering, technical, or maintenance people, they are better placed to quickly and easily detect problems before equipment performance is affected. Thus, the operators become the first line of preventive maintenance in many plants.

However, before involving operators in the routine maintenance of their equipment, certain disciplines *have to* be in place. The first prerequisite is the basic discipline required for effective preventive maintenance and lubrication programs; the second is the discipline of measuring and reporting equipment performance. Without these fundamental disciplines in place, operator involvement becomes fragmented, unstructured, sporadic, and short-lived.

In plants where the disciplines of preventive maintenance, lubrication, and performance measurement are not in place, operators, maintenance, and engineering/technical groups can work together to establish them. This can serve as the beginning of establishing the teamwork among the members of the "natural" work group.

Autonomous maintenance is a structured Japanese approach to operator involvement in maintenance, but it is not the only method for involving operators.

Train Everyone Involved

Education and training have long been recognized as foundations for improved performance in the workplace. In a TPM work culture, education and training are essential for achieving and sustaining new levels of equipment performance and then continually improving upon it. Such education and training spans the spectrum of needs for successful TPM work cultures and typically includes:

- Operating skills and knowledge for operators.
- Maintenance skills and knowledge for maintenance personnel.
- Functional knowledge of process for operations and maintenance personnel.
- Teamwork and leadership skills.
- TPM orientation and methods.
- TPM-specific skills.
- Maintenance skills and knowledge for operators.
- Operating skills and knowledge for maintenance personnel.
- Trainer or on-the-job coach training.

As shown in Figure 10-1, training spans all levels and elements of TPM. It is not a one-time event. Education and training can take many forms, depending on the function and objectives. Commonly associated with TPM are:

- Classroom lecture and demonstration.
- Hands-on, on-the-job training.
- Single-point lessons or job aids.
- Computer interactive training.
- Video-based learning.
- Guided self-study, or
- A combination of the above.

*Worker-Centered Learning** is a good guide to teach the basic maintenance skills to operators and broader, higher level skills to maintenance employees. Mechanics should help develop the training, teach the operators, and determine when they are "qualified." Similarly, maintenance employees skilled in specialty areas such as predictive maintenance should train and qualify their peers.

Design for Early Equipment Management and Maintenance Prevention

In a TPM work culture, when new equipment is specified and ordered, the natural work group is often involved. Those who are charged with designing and ordering new equipment consult with the operators to learn what could be done to make the new piece of equipment easier to operate, easier to inspect, easier to clean. They ask maintenance people what could be done to reduce the time required to maintain the equipment, to make it easier to lubricate, inspect, adjust, repair. Often, representatives of the work group visit the vendor or manufacturer to help make selections or see the equipment in early stages of fabrication and assembly.

When new equipment is specified, its life-cycle cost should be considered, not just the initial purchase price. The costs associated with routine operations, maintenance, power usage, safety and environmental concerns, and disposal have to be taken into account.

Installation of new equipment is preceded by operations, maintenance, and technical training, and the equipment itself arrives with complete documentation, including recommended preventive and predictive maintenance requirements.

FROM CONVENTIONAL MAINTENANCE TO EQUIPMENT MANAGEMENT

Establishing a TPM work culture for a new plant or facility is considerably less difficult than transforming many traditional maintenance practices and organizations. However, moving from traditional maintenance and manufacturing practices to TPM can be made much easier by first understanding all of the concepts and strategies associated with TPM in its purest form and then building on your current work culture and equipment management practices. Look for pockets of TPM practices that may already exist and build on them.

In many businesses the maintenance paradigm or mindset may be the most difficult barrier to overcome in your journey toward a TPM work

---

Worker-Centered Learning: A Union Guide to Workplace Literacy; AFL-CIO HRDI, Washington, DC.

culture. The maintenance paradigm must be shifted to equipment management and reliability improvement as part of the TPM transformation. Look again at the TPM Puzzle in Figure 10-1. Begin at the lower, or foundation, level and move upward. Building a strong maintenance and equipment management foundation is essential for a TPM work culture—and is essential before involving operators in the routine maintenance of their own equipment.

Learn about TPM, what it is, how it works, and why. Visit plants. Attend conferences and seminars. Read books and articles. Learn from the video programs and case examples of TPM elements in U.S. manufacturing. But don't look for a TPM "cookbook." TPM is dynamic and multi-faceted. Successful companies adapt TPM practices to suit the needs of their business and their organizations and people. A growing number of companies embrace the concepts and practices of TPM but never call their initiative TPM. And what happens? Their equipment effectiveness improves and the work force feels energized and renewed by their improvements. That's because TPM is a results-oriented approach rather than an activity-based program for improving manufacturing and maintenance.

Bibliography

Bakerjian, R., ed.; *Tool and Manufacturing Engineers Handbook,* Vol. 7 "Continuous Improvement"; chap. 15, Total Productive Maintenance; Society of Manufacturing Engineers, Dearborn, MI 1993.

Nachi-Fujikoshi Corp., ed.; *Training for TPM*; Productivity Press, Cambridge, MA 1986.

Nakajima, S.; *Introduction to TPM*; Productivity Press, Cambridge, MA 1988.

Ohl, C.J.; Impact of Maintenance on Productivity; SME Technical Paper MM76-318, Society of Manufacturing Engineers, Dearborn, MI 1976.

Osamu, O.; Building a JIT Pyramid; SME Technical Paper MS89-402, Society of Manufacturing Engineers, Dearborn, MI 1989.

Philpot, M.L.; McConkey, J.E.; Vessel, C.M.; and Gyaw, T.A.; The Development of a Generic Total Productive Maintenance (TPM) System; SME Technical Paper MS94-152, Society of Manufacturing Engineers, Dearborn, MI 1994.

Steinbacher, H.R. and N.L.; *TPM for America*; Productivity Press, Cambridge, MA 1993.

Wireman, T.; *Total Productive Maintenance*; Industrial Press, New York, NY 1991.

Wireman, T.; *Computerized Maintenance Management Systems*, 2nd ed.; Industrial Press, New York, NY 1994.

Index

A

Analytical process, 65
Anxiety, 26
AT&T, 86-87, 89-90, 92
Audit performance, 64
Autonomous
   activity(ies), 4, 97
   interaction, 60
   maintenance, 44, 46, 93-94, 96, 100, 107
   teams, 44, 86
Average time between equipment failures, 86
Average time to make repairs, 86

B

Baseline, 33, 59
Baselining, 18, 26, 27
Benchmark(ing), 27, 64
Briggs & Stratton, 47-48, 56, 59
Building utilities, 63

C

Cellular manufacturing, 58
Change continuum, 19
Communication(s), 26, 36-37, 52-54, 56-57, 62
   frequency, 45
   networks, 40
Computerized maintenance management system (CMMS), 33, 57, 71, 79-82, 85-86, 89-90, 92
   CMMS maturity, 82, 84
   CMMS modules
      contractor, 82
      equipment, 82
      inventory and stores, 81
      labor/personnel, 81
      PM, 81
      predictive maintenance (PDM), 81
      purchasing, 81
      rebuild, 82
      work order, 82, 84
Conflict resolution, 42
Continuous improvement, 4, 8-10, 33, 51, 55, 58, 79, 90, 99
Continuous process improvement, 98
Coolant analysis, 61, 77
Critical path machines, 71
Culture, 9, 17, 36, 40, 42, 51, 55, 57-58, 60, 79

D

Decision making, 21-22, 37, 42
Departmental support, 23
Design cycle time, 11-12, 14
Diagnostic
   evaluation, 66
   tools, 73
Documentation, 27

E

Education, 3, 42, 45, 107
Electronic time independent system (ETIS), 92
Elements of TPM, 26
Employee input, 23
Employee involvement (see also total employee involvement), 55, 58
Empowered workers, 20
Empowerment, 15, 17-18, 36, 40, 44, 93-94, 99
Engineering analysis, 33

113

Ensured capacity, 23
Equipment
  analytical support, 73
  capability, 48
  capacity, 48
  critical path, 63-64, 76
  effectiveness, 2, 8, 12, 18, 25-27, 41, 44, 46, 101, 104-105, 109
  -focused work group, 106
  information management, 80
  inventory, 62
  maintenance management, 59
  management, 48, 101-103, 108-109
  management strategy, 2, 101
  -related losses, 105-106
  reliability, 48
  standardization, 10
  support, 63, 76
Equipment management information system (EMIS), 82, 85-86

F

Facilitator(s), 37, 58, 99
Facility evaluation, 62
Failure analysis, 33
Failure mode analysis (FMA), 90
Five Pillars of TPM, 2, 103

G

Generic code, 75, 77
Goal(s), 9, 23, 35, 37, 47-48, 52-53, 98
  long-term, 64
  mid-range, 64
  strategic, 42
  team, 48

H

Harley-Davidson, 23, 33-34, 61, 77-78

I

Infrared
  analysis, 67, 69
  temperature monitoring, 7
Innovation, 21

J

Just in time (JIT), 3, 16, 34, 58, 63, 100
2LAI Maintenance Systems, 41
Life-cycle costs, 3
Lifetime maintenance program, 23
Lube codes, 77
Lubrication, 74-77, 93-94, 98-99, 106

M

Magnavox, 35, 39, 93, 96-98, 100
Maintenance and Calibration Control System (MACCS), 86-91
  MACCS modules
    breakdown records, 87
    equipment spare parts inventory, 89
    facility inventories, 87
    maintenance schedules, 87
    maintenance text, 88
    preventive maintenance, 89
    work orders, 88
Maintenance continuum, 6
Maintenance information tracking system, 85
Mean time between failures, 86
Mean time to repair, 86
Mega route, 72
Mission, 21, 35-36, 39, 47, 52-53, 59-60
  measurement data, 39
  statement, 47-48

N

Natural work group(s), 107-108
  equipment-focused, 104
Needs analysis, 45
Nondestructive testing, 7
Notepad, 83-84

O

Objectives, 9, 36, 47-50, 52, 59
Oil (sampling) analysis, 61, 67, 69, 77

Operator-based routines, 61
Operator involvement (see also total employee involvement), 61
Operator/machine relationship, 95
Organizational maturity, 82-84
Overall equipment effectiveness (OEE), 2, 11, 27, 86
   reliability, 63
Ownership, 21, 36, 54, 93, 95, 99, 101, 104, 106

P

Participative climate, 19
Personalities, 65
Pictorials, 62
Pilot
   area, 25-26, 33-34, 54
   equipment, 18, 27
   project, 63-64, 73
Piloting, 27
Planned maintenance, 68-69, 105
PM-Team(s), 49, 54, 56-57
Predictive
   maintenance, 10, 24, 27, 73, 75, 106
   techniques, 61
   monitoring, 8
   technologies, 59
Premature failure, 76
Preventive
   maintenance (PM), 1, 6-7, 10, 24, 27, 57, 61, 64, 66, 74, 84, 86-87, 106
   time-based, 87
Proactive
   maintenance, 79
   technologies, 59
Problem solving, 21, 42, 95
Product design, 46
Productive maintenance, 1-2
Productivity pyramid, 16

Q

Quality circles, 104
Quality continuum, 4-5

R

Reactive maintenance, 34, 59, 68-70, 74-75, 83

Records control, 71
Recycling, 75
Reliability centered maintenance, 2, 106
Reservoir management, 77
Reuse, 76
Reward system, 56
Risk taking, 20
Role(s), 20, 36, 38

S

Scheduled maintenance (see also preventive maintenance), 69, 106
Shop-floor losses, 2
Small group activities, 23-24, 104
Sonics, 7
Spare parts, 106
Spectrographic oil analysis, 7
Stability, 21
Standards, 33, 98-99
Statistical techniques, 6-7
Status indicators, 76
Strategic Work Systems, 101-102
Surge comparison testing, 69

T

Target areas (see also pilot areas), 54-55
Task(s)
   accountability, 70
   frequency of, 68
   orders (see also work orders), 72, 78
   ownership, 69
   PM, 69, 72
   patterns of, 72
   predictive, 70
   safety, 70
Taylorism, 19
Team, 34-37, 60
   building, 26, 36-37, 42, 96
   coordinating, 43
   cross-functional, 8, 90, 104
   development, 36
   facilitator, 38
   flexible, 39
   FMA, 90
   goals, 97
   leader, 38

Team (cont.)
   member(s), 37-39, 46, 55-56, 97-98
   recorder, 38-39
   self-directing work, 58
   spirit, 20
   template, 39
Team-oriented universe, 37
Teamwork, 20, 35, 37, 56, 60, 93
   equipment-focused, 105
   training, 96
Thermography, 61, 77
Three Concepts of TPM, 103, 105
Timken Company, The, 1, 15, 17
Total employee involvement (TEI), 3, 16, 33
Total quality, 104
   management (TQM), 1, 7, 16, 33
   programs, 3
TPM Puzzle, 102, 109
TPM readiness assessment, 103
Tracking, 71
Training, 3, 24, 33, 41, 43-44, 86, 95, 107
   curriculum, 53
   determining needs and objectives, 45
   formal, 95
   leadership, 44
   needs, 45-46
   on-the-job, 107
   plan, 46
   records, 81
   safety, 96
   session, 46
   skills, 8
   strategy, 42, 108
   technical, 18
   team-building, 8, 18

Trend analysis, 67
Trending, 8
Trust, 21, 35-36, 52-53
   levels, 40

U

Ultrasound analysis, 69
Union, 60
   buy-in, 50
   environment, 47
   involvement, 49-50
   labor, 35-36
   leadership, 55, 103
   management, 49, 56
   officials, 50
   representation, 51

V

Vibration analysis, 7, 61, 67, 69, 77, 81
Vision, 19, 35, 47, 49-50, 52-53, 59
Vision 2000, 15-16

W

Walking route, 71-72
Work order(s), 57, 90, 92, 98

Z

Zero
   breakdowns, 2
   defects, 2, 97
   errors, 54, 58, 62, 76-77
   omissions, 54, 58, 62, 76-77
   scrap, 97